29th Workshop on Recent Developments in Computer Simulation Studies in Condensed Matter Physics 2016

Journal of Physics: Conference Series Volume 750

Athens, Greece, USA
22 – 26 February 2016

Editors:

D.P. Landau
M. Bachmann
S.P. Lewis

ISBN: 978-1-5108-2997-8
ISSN: 1742-6588

Printed from e-media with permission by:

Curran Associates, Inc.
57 Morehouse Lane
Red Hook, NY 12571

Printed by Curran Associates, Inc. (2016)

For permission requests, please contact the Institute of Physics
at the address below.

Institute of Physics
Dirac House, Temple Back
Bristol BS1 6BE UK

Phone: 44 1 17 929 7481
Fax: 44 1 17 920 0979

techtracking@iop.org

Additional copies of this publication are available from:

Curran Associates, Inc.
57 Morehouse Lane
Red Hook, NY 12571 USA
Phone: 845-758-0400
Fax: 845-758-2633
Email: curran@proceedings.com
Web: www.proceedings.com

Table of contents

Volume 750

29th Workshop on Recent Developments in Computer Simulation Studies in Condensed Matter Physics

22–26 February 2016, Georgia, USA

Accepted papers received: 2 September 2016
Published online: 16 September 2016

Preface

Papers

Preface

Thirty years ago, because of the dramatic increase in the power and utility of computer simulations, The University of Georgia formed the first institutional unit devoted to the application of simulations in research and teaching: The Center for Simulational Physics. Then, as the international simulations community expanded further, we sensed the need for a meeting place for both experienced simulators and newcomers to discuss inventive algorithms and recent results in an environment that promoted lively discussion. As a consequence, the Center for Simulational Physics established an annual workshop series on Recent Developments in Computer Simulation Studies in Condensed Matter Physics. This year's highly interactive workshop was the 29th in the series marking our efforts to promote high quality research in simulational physics. The continued interest shown by the scientific community amply demonstrates the useful purpose that these meetings have served.

The latest workshop was held at The University of Georgia from February 22-26, 2016. It served to mark the 30th Anniversary of the founding of the Center for Simulational Physics. In addition, during this Workshop we celebrated the 60th birthday of our esteemed colleague Prof. H.-Bernd Schüttler. Bernd has not only contributed to the understanding of strongly correlated electron system, but has made seminal contributions to systems biology through the introduction of modern methods of computational physics. These Proceedings provide a "status report" on a number of important topics. This on-line "volume" is published with the goal of timely dissemination of the material to a wider audience.

This program was supported in part by the President's Venture Fund through the generous gifts of the University of Georgia Partners and other donors. We also wish to offer thanks to the Office of the Vice-President for Research, the Franklin College of Arts and Sciences, and the IBM Corporation for partial support of this year's workshop.

These Proceedings contain both invited papers and contributed presentations on problems in both classical and quantum condensed matter physics. As usual, topics ranged from hard and soft condensed matter to biologically inspired problems and purely methodological advances. While familiar topics like phase transitions were still on display, the trends in biophysics, dynamical behavior and complex systems demonstrated the continuing progression in the focus of computational condensed matter physics. We hope that readers will benefit from specialized results as well as profit from exposure to new algorithms, methods of analysis, and conceptual developments.

Athens, GA, U.S.A.
April, 2016

D. P. Landau
M. Bachmann
S. P. Lewis

Peer review statement

All papers published in this volume of *Journal of Physics: Conference Series* have been peer reviewed through processes administered by the proceedings Editors. Reviews were conducted by expert referees to the professional and scientific standards expected of a proceedings journal published by IOP Publishing.

Measuring synchronization of stochastic oscillators in biology

Z Deng[1], S Arsenault[2], L Mao[2], and J. Arnold[3]

[1]College of Engineering, University of Georgia
[2]Institute of Bioinformatics and Entomology Department, University of Georgia

[3]Genetics Department, University of Georgia

arnold@uga.edu

Abstract. A fundamental problem in physics is measuring and modeling the synchronization of coupled stochastic oscillators. The problem is relatively recent in biology, where it has become possible to measure stochastic oscillators in single cells. A variety of synchronization measures have been proposed to describe a field of coupled stochastic oscillators. We introduce a synchronization measure new to this problem (but old to Genetics) called the intraclass correlation (ICC). The ICC is simple to interpret and has a statistical framework for inference. We illustrate ICC behaviour in the Kuramoto phase-locking model and on a field of over 25,000 oscillators in single cells measured every half-hour over a ten day interval.

1. Introduction

The following paper is dedicated to H.-Bernd Schuttler on his 60th birthday in honor of his contributions in systems biology drawn from statistical physics(*1-4*). His introduction of ensemble methods into systems biology constituted a sea change for the field, one now that many of us take for granted. The focus today is on a new problem on which Bernd is working that is central to physics and more recently in biology, namely measuring and modeling the synchronization of coupled stochastic oscillators.

The problem arises in physics in the synchronization of lasers, Josephson-Junction arrays, relativistic magnetrons, and the coupled signals of two interferometers measuring gravitational waves (*5, 6*). Only recently has the problem come to the attention of biologists, particularly in the context of circadian rhythms(*7*). With the advent of new technologies for measuring the clocks in single cells(*8*), it is now possible to make progress on how the clocks of individual cells become synchronized. Being able to measure the clocks of individual cells fundamentally changes the scale of measurement usually performed on mass cultures of 10^7 cells. Because of the change in scale, new behaviors are evident that demand new models for the stochastic oscillations of single cells(*9*). Many of the new stochastic models describing the behavior of individual cells have been have been drawn from physics(*10*).

The problem to be described here is simple. Each cell has its own oscillator, and the average behavior of these oscillators has been successfully described with ensemble methods applied to deterministic models at the level of 10^7 cells (*2, 3*). The challenge today is to measure individually each of the oscillators of >25,000 cells over a ten day period and determine how the cells synchronize over time. In contrast to earlier experiments, these new measurements are done at the single cell level. The data are the trajectories of each cell's oscillator over a ten day window measured every half hour (h). The data for one time point are shown in Figure 1 for one field of view and resemble a star field observed at one of successive time points. All of the data at one time point is the union of many fields of view (~30). We will introduce measures of synchronization between the oscillators (stars) drawn from physics and genetics. We will illustrate their behavior in a physical model. Then we will examine these synchronization measures and their behavior on >25,000 individual cells over a 10 day window.

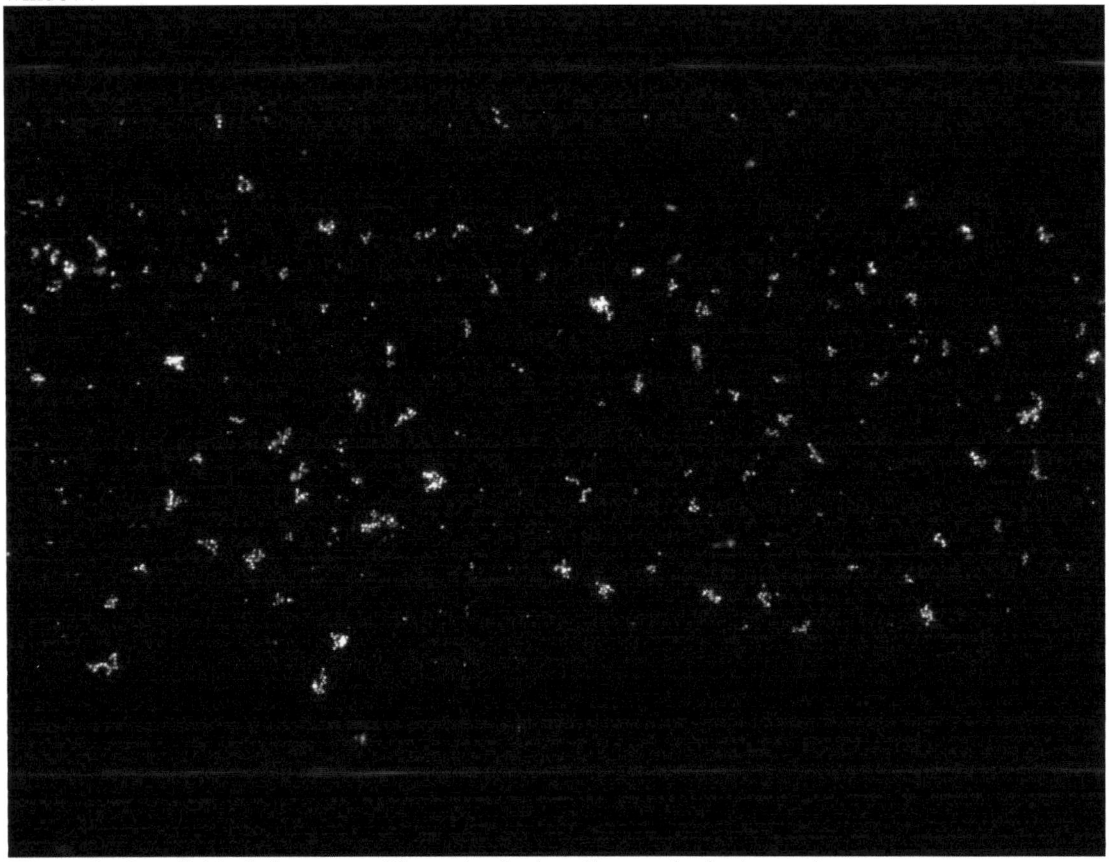

Figure 1. One field of view on oscillators in single cells at a single time point.

2. Synchronization measures

A good synchronization measure should have a number of properties. One, it should capture the similarity between oscillator trajectories as summarized in their phase, amplitude, and period. Second, any measurement of similarity should be done relative to the variability in the measurements between replicates. This will allow an assessment of significance of trajectory similarities found. Three, the measure should have a simple interpretation. Finally, the measure should have well known statistical properties to facilitate inference about synchronization of stochastic oscillators.

A variety of synchronization measures have been introduced to describe the coupled behavior of stochastic oscillators in physics. These include the cross-correlation between cellular trajectories (with respect to frequency), mutual information shared between trajectories, the phase synchronization from either the Hilbert or Wavelet transform of individual cellular trajectories, and an index based on the circular variance known as the phase-locking value(*11*).

Of these methods, we chose to utilize the phase synchronization statistic (or phase locking index) extracted using the Hilbert Transform. By taking the Hilbert Transform of the data

$$\tilde{x}(t) = \frac{1}{\pi} * p.v. \int_{-\infty}^{+\infty} \frac{x(t')}{t - t'} dt'$$

It is possible to define the analytic signal

$$Z_x(t) = x(t) + i\tilde{x}(t) = A_x^H(t) e^{i\varphi_x^H(t)}$$

and consequently extract the phase with respect to time.

$$\varphi_x^H(t) = arctan \frac{\tilde{x}(t)}{x(t)}$$

An index of phase locking can then be calculated to compare the synchronization of two signals.

$$\gamma = \left| \frac{1}{N} \sum_{j=1}^{N} e^{i[\varphi_x(t_j) - \varphi_y(t_j)]} \right|$$

One additional measure has been recently introduced for the study of biological oscillators(*12*). Garcio-Ojalvo et al.(*12*) have suggested one measure of synchronization (R_G), the variance in the measured mean of the oscillator X_{ij} along a trajectory (in time) divided the variance in X_{ij} of an individual trajectory:

$$R_G = <[(n^{-1}\Sigma \bullet X_{ij})^2> - <(n^{-1}\Sigma \bullet X_{ij})>^2]/(<[X_{ij}]^2> - <[X_{ij}]>^2), \quad (1)$$

where n is the number of cell trajectories observed. If the cells are out of phase, then the numerator of R_G will be small relative to the denominator due to phase cancellation. On the other hand, a high degree of synchronization will essentially reduce observations on n cells to the same as looking at one cell, and the measure R will then be near 1. This measure does have then a simple interpretation.

There is also one time-tested measure of similarity that has been used in genetics for over 150 years, the intraclass correlation(*13*). Let us suppose the measurements on the state of oscillator i within the jth neighborhood at one time point are denoted by X_{ij}. A neighborhood is collection of cells that are neighbors. Only oscillators within a neighborhood can be coupled. We observe each cell within a neighborhood every half hour over a ten day period. We can think of neighborhoods of the same size a as replicate experiments.

The measurements on the cells are assumed to have the following covariance structure

$$\begin{aligned} COV(X_{ij}, X_{kl}) &= 0, \quad i \neq k \text{ (different neighborhoods)} \\ &= \rho\sigma \quad i = k, j \neq l \text{ (different cells within a neighborhood)} \\ &= \sigma \quad i = k, j = l \text{ (same cell within a neighborhood)} \end{aligned} \quad (2)$$

Stochastic oscillators within a neighborhood under this model can communicate and are coupled. The parameter ρ is the intraclass correlation and is a direct measure of the coupling of oscillators.

The intraclass correlation has a distinguished history in genetics(*13*) as a heritability measure (in which neighborhoods are families) and two simple interpretations. Firstly, the intraclass correlation is the correlation between measurements on different cells within the same neighborhood. Second, the intraclass correlation can also be interpreted as the fraction of variation in the oscillator explained within neighborhoods relative to that between neighborhoods. In this way the intraclass correlation has a built-in internal control to compare the communication between oscillators within neighborhoods relative to the variance in oscillators across replicate neighborhoods of the same size (a). We will denote the sample estimate of the intraclass correlation by R. This estimate will depend not only on the size of the neighborhoods (a), but will also vary with time. So, the intraclass correlation should be thought of as a synchronization surface varying both in time and neighborhood size. In other words, there will be trios, quartets, and quintets all singing different pieces simultaneously and over time, and these ensembles of musicians will be replicated.

There is a substantial body of theory for inference about the intraclass correlation(*13*). The total variation in oscillator behavior (X_{ij}) at any particular time point can be partitioned into a between

Journal of Physics: Conference Series **750** (2016) 012001 doi:10.1088/1742-6596/750/1/012001

neighborhood sum of squares (SS_B) and a within neighborhood sum of squares (SS_W) in Table 1 such that the total sum of squares $SS_T = SS_B + SS_W$. The ratio of Estimated Mean Squares (EMS) provide a test of the significance of the intraclass correlation ρ known as an F-test.

Table 1. Analysis of Variance (ANOVA) of oscillators between and within neighborhoods is used to estimate the intraclass correlation ρ. The model in (2) is a balanced variance components model with cells grouped by neighborhoods(*13*). The number of cells in a neighborhood is denoted by a; the number of neighborhoods is denoted by n. The variance in a single oscillator measurement is σ and the intraclass correlation is ρ The intraclass correlation is a measure of the fraction of total variation (SS_T) within neighborhoods as well as the correlation between oscillator measurements on different cells within the same neighborhood. The quantities, $XBAR_i$ and XBAR, are sample means over neighborhood i and over cells and neighborhoods, respectively. The estimated mean square (EMS) is a sum of squares (SS) divided by its degrees of freedom. The EMS estimates its expectation in the last column. Setting the EMS equal to their expectations allows the solution for ρ.

Source	degrees of freedom (df)	Sums of Squares (SS)	Estimated Mean Square (EMS)	Expectation of EMS
Between neighborhoods	n-1	$SS_B = a\ \Sigma_i(XBAR_i - XBAR)^2$	$SS_B/(n-1)$	$\sigma^2 + (a-1)\rho\sigma^2$
Within neighborhoods	n(a-1)	$SS_W = \Sigma_i\Sigma_j(X_{ij} - XBAR_i)^2$	$SS_W/n(a-1)$	$(1-\rho)\sigma^2$
Corrected total	na-1	$SS_T = \Sigma_i\Sigma_j(X_{ij} - XBAR)^2$		

From the Analysis of Variance (ANOVA) Table 1, an estimator of the intraclass correlation can be constructed by setting the Estimated Mean Squares (EMS) equal to their expectations. The result is two equations in two unknowns, ρ and σ Solving these two equations yields the estimator R:

$$R = (EMS_B - EMS_W)/(EMS_B + (a-1)EMS_W) \quad (3)$$

The estimator R defines a surface describing the synchronization of the oscillators in the system as a function of time (t) and neighborhood size (a). This synchronization measure has the four properties desired of a synchronization measure outlined earlier.

3. Kuramoto phase locking model

Kreuz et al.(*11*) concluded from a comparative analysis of varied synchronization measures in different physical systems that no one synchronization measure was to be preferred in different physical systems, but the modeling context determined the appropriate synchronization measure to be used in a particular physical system. One of the classic physical models of phase locking is the Kuramoto model(*14*). If the oscillators within cells are constrained to a ring indexed by r, then neighboring oscillators communicate. The model is defined by the following equations. For $s \in [1, N]$,

$$\dot{\varphi}_s = \omega_s + \frac{K}{N} * \sum_{r=1}^{N} \sin(\varphi_r - \varphi_s)$$

The parameters used here are $\omega_s = \frac{2*\pi}{20}$, K = 0.05. The mapping onto the fluorescent measurements is $X = \sin\varphi$.

In Figure 2 is displayed the synchronization surface for the Kuramoto phase-locking model using the intraclass correlation R in Eqn. (3). The synchronization surface (Figure 2) has substantial structure. The measure increases with time for different neighborhood sizes. There is also an alternating structure to the surface, where even numbers of neighbors have ridges, while odd numbers of neighbors have valleys. There is more synchronization with an even number of singers, such as a quartet, than for example, a trio.

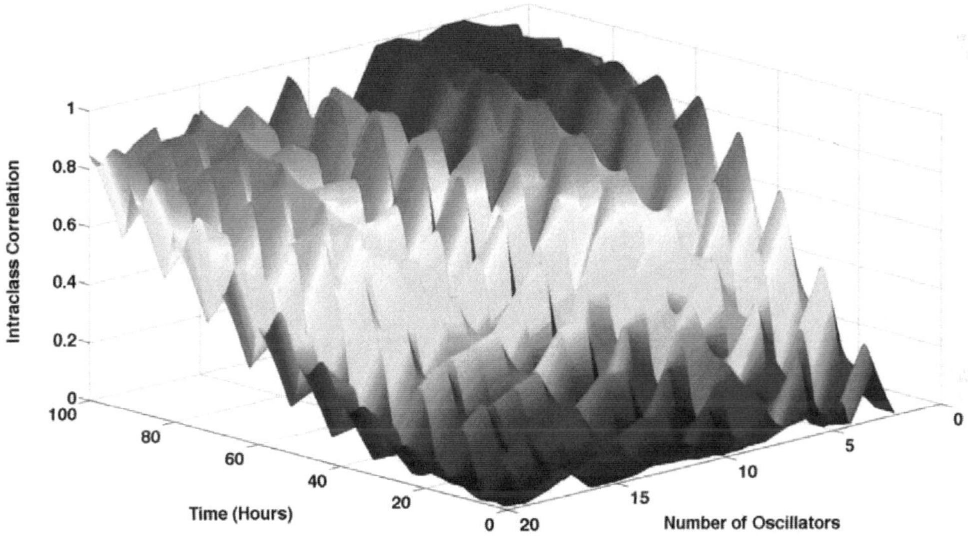

Figure 2. Synchronization surface for the Kuramoto phase-locking model has substantial structure as a function of time (t) and neighborhood size (a).

Journal of Physics: Conference Series **750** (2016) 012001 doi:10.1088/1742-6596/750/1/012001

4. Measuring synchronization

Synchronization was measured on a "star field" of over 25,000 cells in neighborhoods varying in size from 2 to 47 cells (Table 2).

Table 2. Number of Replicate Neighborhoods with Neighborhood size.

Number of Neighborhoods (n)	Number of cells per neighborhood (a)
670	2
436	3
422	4
336	5
310	6
281	7
244	8
224	9
186	10
162	11
120	12
88	13
95	14
66	15
53	16
44	17
29	18
29	19
16	20
13	21
13	22
6	23
14	24
7	25
4	26
2	27
2	29
3	30
4	31
2	32
2	46
2	47

We computed the two older measures of synchronization, the phase of the Hilbert Transform and the R_G measure in (1) for neighborhoods of size 4 (Figure 3) using a one day non-overlapping window. We can see that both measures are roughly increasing with time. These measures only capture one aspect of the similarities of oscillator trajectories, their phase, and do not factor in the between replicate variability across neighborhoods.

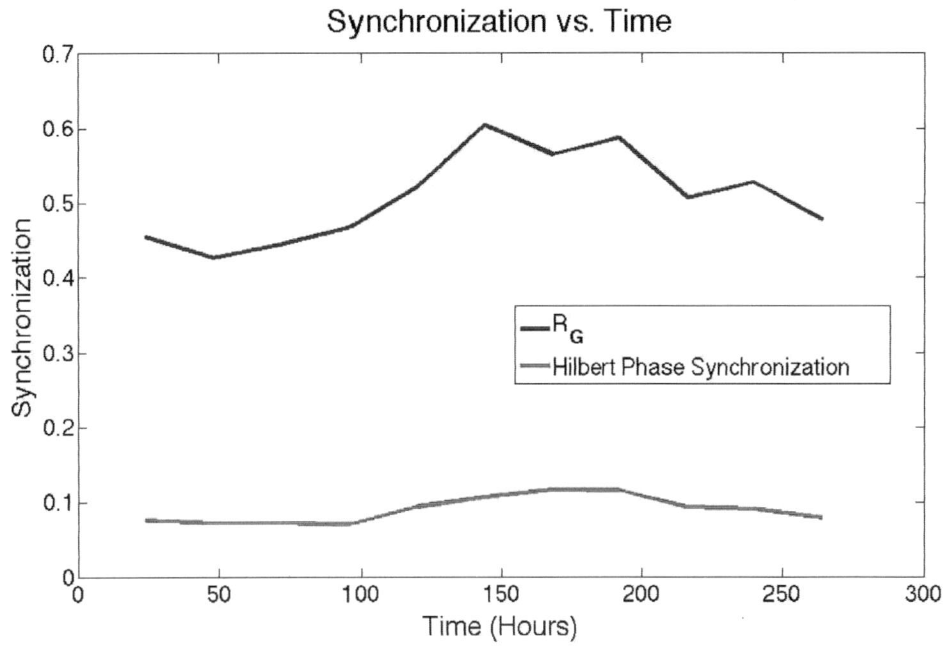

Figure 3. Synchronization measures increase with time for a neighborhood of size 4. Measures graphed are the phase extracted from the Hilbert transform of each trajectory and the synchronization measure R_G in (2).

One of the advantages of the intraclass correlation as a measure of similarity between trajectories is that it also measures the fraction of variation explained by communication within neighborhoods. For example, the intraclass correlation (Figure 5A below) asymptotes at ~50% for most neighborhoods, implying that ~50% of the oscillator variation can be attributed to that within neighborhoods and only 50% to variation between neighborhoods. This begs the question of why oscillator behavior is so similar within neighborhoods.

In particular with the inference tools available about the intraclass correlation(*13*), we can ask questions about the significance of the synchronization. For example, is there any evidence of synchronization when there are only 2 oscillators in a neighborhood (Figure 4)? We constructed the ANOVA tables (from Table 1) for time point 0 and time point 282 hours (Tables 3 and 4). The synchronization was highly significant with R = 0.35 at time 0 and R = 0.40 at time = 282 h. All synchronization values (Figure 4) are significant at all time points even after a multiple test correction.

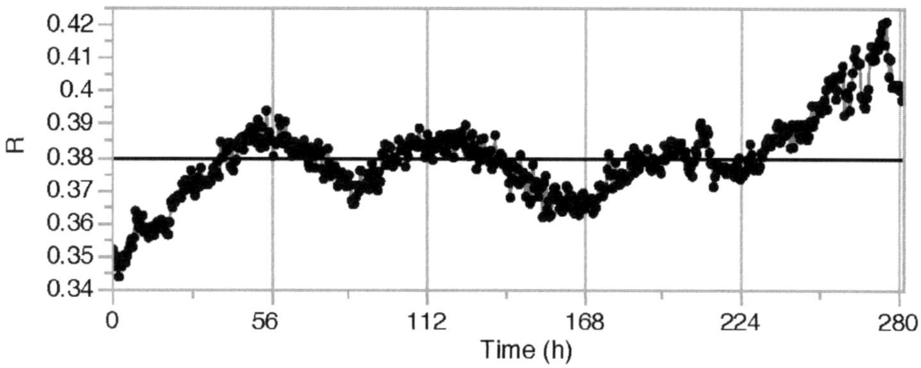

Figure 4. The intraclass correlation R is significant for all time points for neighborhoods with 2 oscillators even after a multiple test correction. This is a slice through Figure 5A.

Table 3. Analysis of Variance (ANOVA) of oscillators between and within neighborhoods is used to test the significance of the intraclass correlation ρ for neighborhoods of size 2 at time point 0. The method of computation of the ANOVA is given in Table 1. The value is significant at least the 0.001 level (***).

Source	degrees of freedom (df)	Sums of Squares (SS)	Estimated Mean Square (EMS)	Expectation of EMS	F
Between neighborhoods	670-1	84639.67452	126.5167033	$\sigma^2 + (a-1)\rho\sigma^2$	
Within neighborhoods	670(2-1)	40642.82316	60.66093009	$(1-\rho)\sigma^2$	0.48***
Corrected total	(670)(2)-1	125282.4977			

Table 4. Analysis of Variance (ANOVA) of oscillators between and within neighborhoods is used to test the significance of the intraclass correlation ρ for neighborhoods of size 2 at time point 282 h. The method of computation of the ANOVA is given in Table 1.

Source	degrees of freedom (df)	Sums of Squares (SS)	Estimated Mean Square (EMS)	Expectation of EMS	F
Between neighborhoods	670-1	26295.44231	39.30559389	$\sigma^2 + (a-1)\rho\sigma^2$	
Within neighborhoods	670(2-1)	11364.75646	16.96232307	$(1-\rho)\sigma^2$	0.43***
Corrected total	(670)(2)-1	37660.19877			

As an alternative, we constructed the synchronization surface as described in Table 1 and (3) for varying time and neighborhood size (Figure 5A) We can see that the synchronization surface highly resembles that produced by the Kuramoto phase-locking model (Figure 2). The ridges and valleys are evident. The ridges in synchronization become higher as time advances. As a comparison we also constructed R_G (Figure 5B). There appears to be an increase of synchronization with time and drop off with neighborhood size.

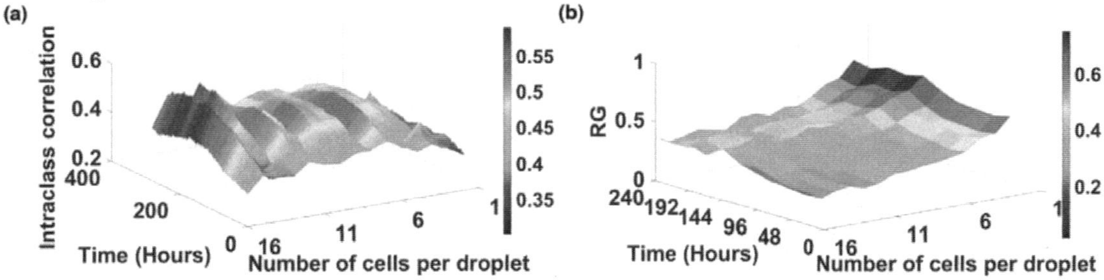

Figure 5. The measured synchronization surface of >25,000 oscillators resembles that of the Kuramoto phase-locking model. (A) Synchronization surface for R in (3). (B) Synchronization surface for R_G in (1) using a one day window.

5. Discussion

The degree of synchronization not only varies with time, but also with the number of neighbors. There is a social environment that affects the behaviour of oscillators. There is structure to the synchronization surface as well (Figure 5). Whether there are an odd or even number of neighbors has an impact on the extent of synchronization. The synchronization will be hard to measure unless the observer controls for neighborhood size (Figure 5).

The intraclass correlation in (3) is readily interpreted in two ways. One, the synchronization measure R asymptoting in time at ~50% is the correlation in individual trajectories within a neighborhood. The final correlation between trajectories is quite high relative to the variation between neighborhoods. The ~50% is also the fraction of variation in oscillator trajectories captured within neighborhoods measured relative to the variation between replicate neighborhoods. The measure also allowed us to test how synchronization arises (Figure 4).

As synchronization takes place with time, there is not a sharp transition as in many physical and chemical systems(*15*). The synchronization is gradual as captured in the intraclass correlation. There is not an abrupt transition to synchronization as seen in some physical systems. It is possible that there is another state of the system as determined by cell density and the degree of coupling of cell oscillators that a sharp phase transition would occur(*15*). The neighborhoods, such as a =2 or the large neighborhoods (a=42), display non-monotone synchronization because of noise or the small number of replicate neighborhoods to average over to produce the surface value. For the other neighborhoods, the synchronization surface is increasing in time.

The synchronization surface (Figure 5) is also a rich source for testing various physical and biological models of communication. The Kuramoto phase locking model reproduces the structure in the synchronization surface, implying that there is a local phase-locking mechanism in operation between oscillators in cells within neighborhoods. Further information could be obtained by ensemble fitting of the Kuramoto model to the trajectories of individual oscillators(*2, 16*).

The problem of measuring synchronization in physics and biology is a general one. The collective behavior of individual actors and their synchronization is observed in a variety of biological and physical systems. One recent example is the highly synchronized marching of locusts or the behavior of human crowds(*16, 17*). The study of synchronization in physical and biological problems will continue to be challenging experimentally and discerning its underlying mechanism(s), a fundamental problem in both physics and biology.

References

1. D. Battogtokh, D. K. Asch, M. E. Case, J. Arnold, H. B. Schuttler, An ensemble method for identifying regulatory circuits with special reference to the *qa* gene cluster of *Neurospora crassa. Proceedings of the National Academy of Sciences of the United States of America* **99**, 16904-16909 (2002).
2. Y. Yu *et al.*, A genetic network for the clock of *Neurospora crassa. Proceedings of the National Academy of Sciences of the United States of America* **104**, 2809-2814 (2007).
3. W. Dong *et al.*, Systems biology of the clock in *Neurospora crassa. PloS one* **3**, e3105 (2008).
4. A. Al-Omari *et al.*, Discovering Regulatory Network Topologies Using Ensemble Methods on GPGPUs With Special Reference to the Biological Clock of *Neurospora crassa. IEEE Access* **3**, 27-42 (2015).
5. S. H. Strogatz, Exploring complex networks. *Nature* **410**, 268-276 (2001).
6. B. Abbott *et al.*, Observation of Gravitational Waves from a Binary Black Hole Merger. *Physical Review Letters* **116**, 061102 (2016).
7. A. T. Winfree, On emerging coherence. *Science* **298**, 2336-2337 (2002).
8. O. Mondragon-Palomino, T. Danino, J. Selimkhanov, L. Tsimring, J. Hasty, Entrainment of a population of synthetic genetic oscillators. *Science* **333**, 1315-1319 (2011).
9. Z. Deng *et al.*, Single cell measurements on the biological clock by microfluidics. *Proc. of the 18nth International Conference on Miniaturized Systems for Chemistry and Life Sciences (MicroTAS). San Antonio, TX* **18**, 881-883 (2014).
10. J. D. Challenger, A. J. McKane, Synchronization of stochastic oscillators in biochemical systems. *Physical Review E* **88**, 012107 (2013).
11. T. Kreuz *et al.*, Measuring synchronization in coupled model systems: A comparison of different approaches. *Physica D: Nonlinear Phenomena* **225**, 29-42 (2007).
12. J. Garcia-Ojalvo, M. B. Elowitz, S. H. Strogatz, Modeling a synthetic multicellular clock: repressilators coupled by quorum sensing. *Proceedings of the National Academy of Sciences of the United States of America* **101**, 10955-10960 (2004).
13. S. R. Searle, G. Casella, C. E. McCulloch, *Variance Components*. (John Wiley & Sons, 2009), vol. 391.
14. Y. Kuramoto, *Chemical oscillations, waves, and turbulence*. (Springer Science & Business Media, 2012), vol. 19.
15. A. F. Taylor, M. R. Tinsley, F. Wang, Z. Huang, K. Showalter, Dynamical quorum sensing and synchronization in large populations of chemical oscillators. *Science* **323**, 614-617 (2009).
16. D. J. Sumpter, *Collective animal behavior*. (Princeton University Press, 2010).
17. J. Buhl *et al.*, From disorder to order in marching locusts. *Science* **312**, 1402-1406 (2006).

Flat-histogram methods in quantum Monte Carlo simulations: Application to the t-J model

Nikolaos G. Diamantis[(1)] **and Efstratios Manousakis**[(1,2)]

[(1)]Department of Physics, University of Athens, Panepistimioupolis, Zografos, 157 84 Athens, Greece

[(2)]Department of Physics and National High Magnetic Field Laboratory, Florida State University, Tallahassee, FL 32306-4350, USA

E-mail: manousakis@magnet.fsu.edu

Abstract. We discuss that flat-histogram techniques can be appropriately applied in the sampling of quantum Monte Carlo simulation in order to improve the statistical quality of the results at long imaginary time or low excitation energy. Typical imaginary-time correlation functions calculated in quantum Monte Carlo are subject to exponentially growing errors as the range of imaginary time grows and this smears the information on the low energy excitations. We show that we can extract the low energy physics by modifying the Monte Carlo sampling technique to one in which configurations which contribute to making the histogram of certain quantities flat are promoted. We apply the diagrammatic Monte Carlo (diag-MC) method to the motion of a single hole in the t-J model and we show that the implementation of flat-histogram techniques allows us to calculate the Green's function in a wide range of imaginary-time. In addition, we show that applying the flat-histogram technique alleviates the "sign"-problem associated with the simulation of the single-hole Green's function at long imaginary time.

1. Introduction

The quantum Monte Carlo simulation technique has been very successful when dealing with equilibrium properties of system of particles which do not obey Fermi statistics[1]. This technique cannot be applied directly to real-time dynamics and one resorts in the calculation of correlation functions in imaginary-time. From the behavior of these correlation functions in imaginary-time, ground-state properties can be accurately calculated. In principle, an accurately known correlation function at long imaginary-time scales can yield useful information about the low-lying energy excitations. However, typical imaginary-time correlation functions calculated in quantum Monte Carlo are subject to exponentially growing errors as the range of imaginary-time grows. Here, we show that this problem can be significantly alleviated by applying flat-histogram ideas[2, 3, 4] in the Monte Carlo sampling of configurations which contribute to making the histogram of such correlation functions in imaginary-time flat.

Flat-histogram ideas have been successfully applied to classical simulations of systems undergoing first order phase transitions, systems with rough energy landscapes, etc. The Wang-Landau algorithm[4] (WLA), a particularly useful flat-histogram algorithm, has been applied to the simulation of *equilibrium* statistical mechanical properties of *quantum* systems[5]. Following this work, Gull *et al.*[6] have applied the idea to the continuous-time quantum Monte

Carlo approach to the quantum impurity solver needed for all dynamical mean-field theory applications.

The main idea used in the present paper has been demonstrated[7] in the past using the Fröhligh polaron problem[8] by combining the so-called multicanonical[2] (MUCA) or the Wang-Landau algorithm[4] with the diagrammatic Monte Carlo (diag-MC) method[9, 10]. In the present paper, we illustrate the benefit of applying flat-histogram techniques to the diag-MC sampling[9, 10] in order to reveal the imaginary-time dependence of the single-hole Green's function moving in an antiferromagnetic background as described by the two-dimensional (2D) t-J model[11, 12, 13, 14, 15]. The t-J model is probably the simplest non-trivial quantum many-body model to capture the interplay between charge-carrier motion and antiferromagnetic fluctuations and it is as basic as the Ising model for classical systems. Therefore, we feel that we should not proceed any further with quantum simulation of electronic systems without having a technique which can accurately simulate such a model.

First, we use the diag-MC method in conjunction with the flat-histogram technique to the problem of a single-hole in a modified soluble version of t-J model, where, the diag-MC sampling space is restricted to the diagrams which are summed up by the non-crossing approximation (NCA). This allows us to assess the correct implementation and accuracy of the method and to have an "exact" solution to compare with the results obtained with and without application of the flat-histogram method.

Next, we sample all the diagrams without any restriction which corresponds to the fully linearized version of the t-J model without any further approximation. In this case, the sampling technique should sum up amplitudes which are not positive definite and, thus, they cannot be simply interpreted as probability in the Monte Carlo sampling technique. Namely, we encounter the so-called "sign-problem" in the application of the Monte Carlo method. We show that even in this case, where there is the "sign-problem", we can still extract more accurate results for the imaginary-time dependence of the single-hole spectral function and, thus, more accurate results for the low-lying spectrum of the problem.

We will use two different implementations of the diag-MC method. One in which the imaginary-time dimension is not sampled by the Markov process and it is treated as a fixed parameter in a particular simulation. This requires repetition of the simulation for each one of the different values of imaginary-time needed. In the second implementation the imaginary-time is one additional dimension of the sampling space. While the latter approach may be more efficient, we will also use the former approach to demonstrate a different way to apply the flat-histogram idea in which the distribution of the order n contribution to the perturbation expansion is made flat.

2. The Hamiltonian

We use a simplified version of the 2D t-J model in which the Hamiltonian and the hole-hopping terms are linearized within the spin-wave approximation to obtain a polaron-like Hamiltonian ([11, 13, 14]), i.e.,

$$\hat{H} = -\sum_{\vec{k},\vec{q}} g(\vec{k},\vec{q}) a^{\dagger}_{\vec{k}+\vec{q}} a_{\vec{k}} b_{\vec{q}} + H.c. + \sum_{\vec{k}} \hbar\omega(k) b^{\dagger}_{\vec{k}} b_{\vec{k}}, \tag{1}$$

$$g(\vec{k},\vec{q}) = \frac{4t}{\sqrt{N}} (u_{\vec{q}}\gamma_{\vec{k}-\vec{q}} + v_{\vec{q}}\gamma_{\vec{k}}), \qquad \gamma_{\vec{k}} = -2t(\cos(k_x a) + \cos(k_y a)), \tag{2}$$

where $b^{\dagger}_{\vec{q}}$ is the Bogoliubov spin-wave creation operator, $\omega(k)$ is the spin-wave dispersion of the square lattice quantum antiferromagnet and $a^{\dagger}_{\vec{k}}$ is the hole creation operator. Also the $g(\vec{k},\vec{q})$ is the coupling of the hole to spin waves and $u_{\vec{q}}$ and $v_{\vec{q}}$ are the coefficients of the Bogoliubov transformation. For details of the derivation of this simplified version of the t-J model, which remains a non-trivial quantum many-body problem, as well as for the definitions of the operators and the expression for ω_k, $u_{\vec{q}}$, $v_{\vec{q}}$ and $g(\vec{k},\vec{q})$, the reader is referred to Refs. [11, 13, 14].

In order to demonstrate the importance of applying flat-histogram techniques, we will consider two cases. First, no guidance function is used in the diag-DMC simulation, which we will refer to as "bare" diag-MC, which is the straightforward way to apply the diag-MC. This is not the same as the standard implementation of diag-MC[9, 10] where some different "tricks" are applied to assist the simulation at long imaginary time. The second approach which will discuss is when a flat-histogram technique is implemented in the diag-MC simulation. When we apply any of the flat-histogram techniques in conjunction with diag-MC, we will refer to it as flat-histogram diag-MC.

3. Fixed-time diagrammatic Monte Carlo

The diagrammatic Monte Carlo technique[9, 10] is a Markov process in a space defined by all the terms (or diagrams) which appear in perturbation theory. For example, it samples the terms of the perturbation expansion of the imaginary-time single-particle Green's function, which can be schematically written as follows:

$$G(\tau) = \sum_{n}^{\infty} O_n(\tau), \quad O_n(\tau) = \sum_{\lambda} D_n^{(\lambda)}(\tau) \tag{3}$$

$$D_n^{(\lambda)}(\tau) = \int d\vec{x}_1 d\vec{x}_2 ... d\vec{x}_n S_n^{(\lambda)}(\vec{x}_1, \vec{x}_2, ..., \vec{x}_n, \tau), \tag{4}$$

where O_n represents the sum of all the diagrams of order n, $O_0(\tau) = G^0(\tau)$ is the Green's function in zeroth order, and λ is a variable which labels the contribution $D_n^{(\lambda)}$ of a particular diagram of order n. As the order n of the expansion increases, the number of integration variables increases in a similar manner. Notice that when we refer to the nth order in the case of the t-J model we mean that the number of spin-wave-propagators contained in the diagrams is n; therefore, the order in perturbation expansion is $2n$. In diag-MC the random walk makes a series of transitions between states $\{n, \lambda, \vec{R}\} \rightarrow \{n', \lambda', \vec{R}'\}$, where $\vec{R} = (\vec{x}_1, ..., \vec{x}_n)$. Through such a Markov process the entire series of terms is sampled. This process generates a histogram which represents the number of times $N_n(\tau)$ the order n appeared in the Markov process. Since we can compute a low order diagram analytically, say for example the zeroth order $O_0(\tau)$, the absolute value of all other orders is computed as follows:

$$O_n(\tau) = \frac{N_n(\tau)}{N_0(\tau)} O_0(\tau). \tag{5}$$

3.1. The problem with "bare" application of diag-MC

First, we restrict our QMC computation of $G(\tau)$ to sample the subspace spanned only by the diagrams included in the non-crossing approximation (NCA). We do that because in this case, the NCA diagrams can be summed up "exactly" ([13, 14]) and we can use this "exact" solution to judge the accuracy of our results.

In Fig. 1(a) $O_n(\tau)$ as a function of n is shown for a fixed value of τ as calculated for this soluble model. Notice that the distribution of the order n is Gaussian-like which peaks at a value of $n = n_{max}(\tau)$. Fig. 1(b) shows $O_n(\tau)$ on a logarithmic scale for $\tau = 8$, 12 and 16. Notice that as a function of τ, $n_{max}(\tau)$ grows almost linearly with τ, the value of $O_n(\tau)$ at the maximum grows dramatically with increasing τ. Namely, as τ increases higher and higher order diagrams give the most significant contribution. As a result, for large enough τ, for any given limited number of Monte Carlo steps, the number of walks landing in small values of n becomes very small or non-existent. However, when the number of MC steps which land in the state $n = 0$

Journal of Physics: Conference Series **750** (2016) 012002 doi:10.1088/1742-6596/750/1/012002

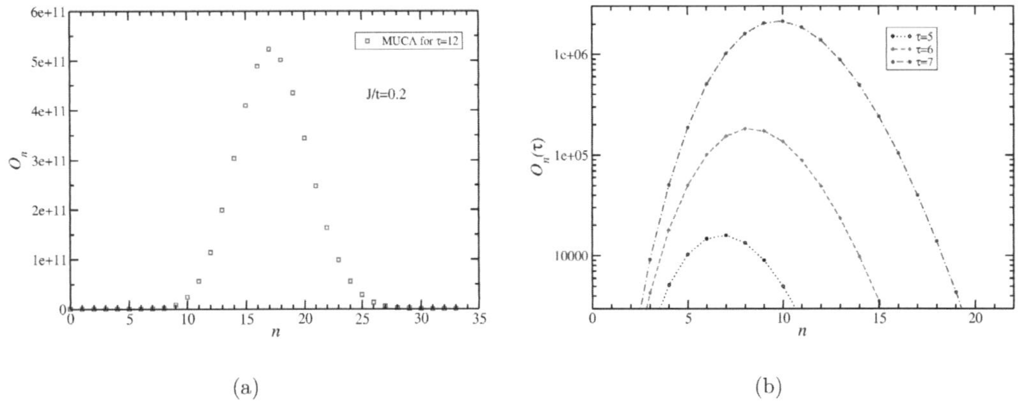

Figure 1. (a) $O_n(\tau)$ as a function of n for fixed τ. (b) $O_n(\tau)$ as a function of n for three different values of τ plotted on a logarithmic scale.

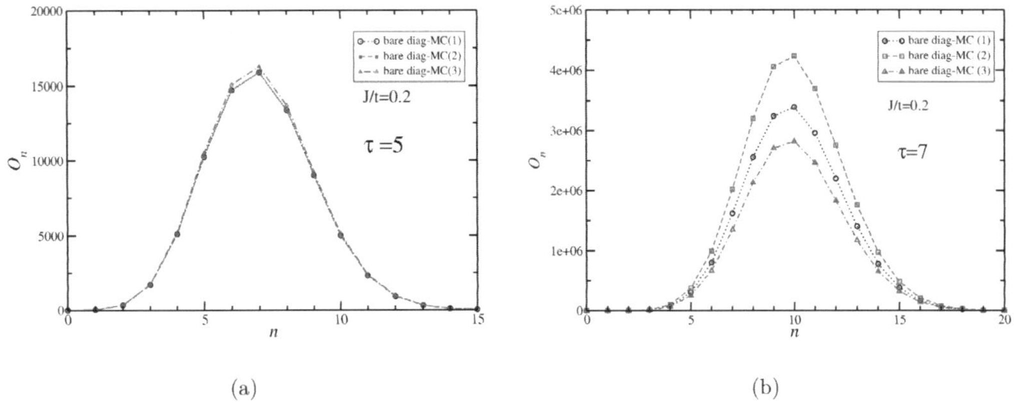

Figure 2. (a) $O_n(\tau)$ as a function of n for $\tau = 5$ as calculated by repeating the diag-MC for 3 different starting configurations. (b) Same as part (a) for $\tau = 7$.

is zero or very small, it leads to a fatal situation in our attempt to calculate the absolute value of $O_n(\tau)$, because this is obtained using the formula (7) and a very small $N_0(\tau)$ implies a large uncertainty in the absolute value of all $O_n(\tau)$.

This is illustrated in Fig. 2 where the calculated $O_n(\tau)$ is shown as a function of n for $\tau = 5$ and $\tau = 7$ as calculated by repeating a diag-MC simulation of 3×10^8 MC steps for 3 different starting configurations. Notice that the statistical fluctuations from one simulation to another affect the values of O_n uniformly for all n by the same fluctuating scale factor, i.e., $1/N_0(\tau)$. If we are to calculate the error from these fluctuations for each value of n, the size of the error bars would be much larger than the size of the fluctuations of the points for successive values of n. Namely, the points which represent $O_n(\tau)$ form a rather smooth curve. This seems unusual given the size of the error bars. This can be explained by the fact that the error is due to the poor estimation of N_0 which propagates through the formula given by Eq. 7. Note that using the same number of MC steps becomes impossible to calculate $O_n(\tau)$ beyond this value of $\tau = 12$ because the ratio $O_{nmax}(\tau)/O_0$ becomes much larger than the number of MC steps.

3.2. Application of the flat-histogram technique

Here, we will solve the problem discussed in the previous subsection by adopting flat-histogram techniques which have been applied in simulations of classical statistical mechanics[2, 4]. We map the particular value of n to the "energy" level in standard flat-histogram methods for classical statistical mechanics and the sum of the terms giving $O_n(\tau)$ to the density of states which corresponds to the corresponding configurations. The flat-histogram method renormalizes the density of states $O_n(\tau)$ for each n by known factors (which can be easily estimated) and, then, samples a more-or-less flat-histogram of such populations.

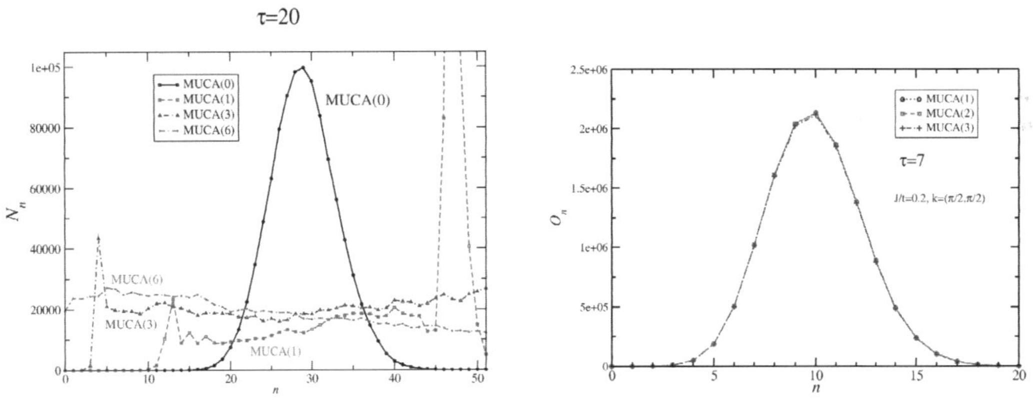

(a) (b)

Figure 3. (a) The evolution of the re-weighted distribution as a function of the multicanonical steps. Notice that the range where the distribution has a significant value expands as we repeat the multicanonical steps and after 6 such iterations (green curve labeled MUCA(6)) it becomes more or less flat in the entire domain. (b) $O_n(\tau)$ as a function of n for $\tau = 7$ as calculated by using multicanonical diag-MC for 3 different starting configurations and by applying the multicanonical method.

Next, we use the idea of the MUCA algorithm[2] as follows: First, for a given fixed value of τ we carry out an initial exploratory run, where we find that the distribution O_n of the values of n peaks at some value of $n = n_{max}$, which depends on the chosen value of τ. The black solid curve in Fig. 3(a) shows the result obtained for the histogram using $M_0 = 10^6$ diag-MC steps. This distribution falls off rapidly for $n > n_{max}$, and, thus, we can determine the maximum value n_c of n visited by the Markov process. We choose a value of m safely greater than n_c, such that the value of O_m is practically zero. Then, we modify the probabilities associated with a particular configuration of the nth order by dividing the original probabilities by a factor $f_n = max(1, N_n)$. Using these modified probabilities we carry out another set of M_0 diag-MC steps which yields a new histogram with populations N'_n shown by the red curve in Fig. 3(a). In the next step, we divide the probabilities associated with a particular configuration of the nth order by the factor $f'_n = f_n * max(1, N'_n)$. Using these modified probabilities we carry out a new set of M_0 diag-MC steps which yields a new histogram with populations N''_n, etc. The blue and green curves in Fig. 3(a) are obtained in the third and sixth iteration of this process. Notice that already at the 6th step the histogram is reasonably flat. When, the histogram becomes more-or-less "flat" at some k^{th} step, we begin a Markov process for a relatively large number of MC steps, by dividing the original probabilities by the factor $f_n^{(k)}$, and, by re-weighting the observables by the biasing factor $f_n^{(k)}$ we determine $O_n(\tau)$ and $G(\tau)$.

Fig. 3(b) shows the results of applying the MUCA algorithm as discussed in the previous paragraph for the same number of MC steps and approximately the same amount of CPU time as in the calculation with the straightforward application of the diag-MC to obtain the curves in Fig. 2(b). Notice the significant reduction of the statistical fluctuations between the three different simulations. Furthermore, the flat-histogram approach allows us to calculate $O_n(\tau)$ for almost any τ, something which is not possible using bare diag-MC.

4. Sampling the imaginary time τ in the diag-MC

Here we discuss the usual implementation of the diag-MC where the imaginary time variable is also a dimension of the sampling space. In this case the random walk makes a series of transitions between states $\{n, \lambda, \vec{R}, i\} \rightarrow \{n', \lambda', \vec{R}', i'\}$, where $\vec{R} = (\vec{x}_1, ..., \vec{x}_n)$, n is the perturbation order, λ is a particular diagram, and i (or i') is the label of a particular imaginary time interval $(\tau_i - \delta\tau/2, \tau_i + \delta\tau/2)$ (where $\delta\tau = \tau_{i+1} - \tau_i$). Namely, in this case we sample the histogram of $G(\tau)$ by including the imaginary time as an extra dimension of the sampling space. In this case, the value $\mathcal{G}_i$ of the histogram $G(\tau_i)$ in the ith τ-interval is found by using the known value of $\mathcal{G}_0$ in the first interval, namely,

$$\mathcal{G}_i = \frac{\mathcal{N}_i}{\mathcal{N}_0}\mathcal{G}_0, \tag{6}$$

where $\mathcal{N}_i$ is the number of occurrences of the MC random walk in the ith interval.

In Fig. 4 we illustrate what the problem is with the bare diag-MC. We first notice that the error in $G(\tau)$ for a given number of MC steps grow exponentially with the maximum value of τ_{max} used in this simulation. The reason is that because of the exponential dependence of $G(\tau)$ on τ itself (See Fig. 4(a)), given a fixed number of MC steps, when τ is sampled, during the Markov process, the number of occurrences $\mathcal{N}_0$ in the $\tau = 0$ interval is exponential small with the value of τ_{max} used. Since the value of $\mathcal{N}_0$ enters in the Eq. 6 for $\mathcal{G}_i$ the fluctuations in $\mathcal{G}_i$ increase exponentially with increasing $\tau_{\max}$ in this case. This is illustrated in Fig. 4(b) where the relative errors in $\mathcal{G}_i$ are given for various values of τ_{max}. For comparison, the relative errors obtained by using the Wang-Landau technique to make the histogram of $\mathcal{G}_i$ flat and for approximately the same amount of CPU time to that used in the bare diag-MC simulation, is also shown in Fig. 4(b). Notice that by increasing $\tau_{\max}$ the error in the bare diag-MC increases more or less uniformly for all values of τ and it is much larger than the error obtained after the application of the WLA.

5. Results for the full t-J model

Here, we present the results for the full t-J model, where we sample all the diagrams using diag-MC. First, we will keep the imaginary time fixed and in the second part we will present the case when τ is also sampled.

5.1. Fixed-time diagrammatic Monte Carlo

First, we note that some terms in a given order O_n have a positive contribution, while some other terms have a negative contribution. We separate these contributions to O_n^+ and $-O_n^-$ such that $O_n = O_n^+ - O_n^-$ where both $O_n^{\pm}$ are positive.

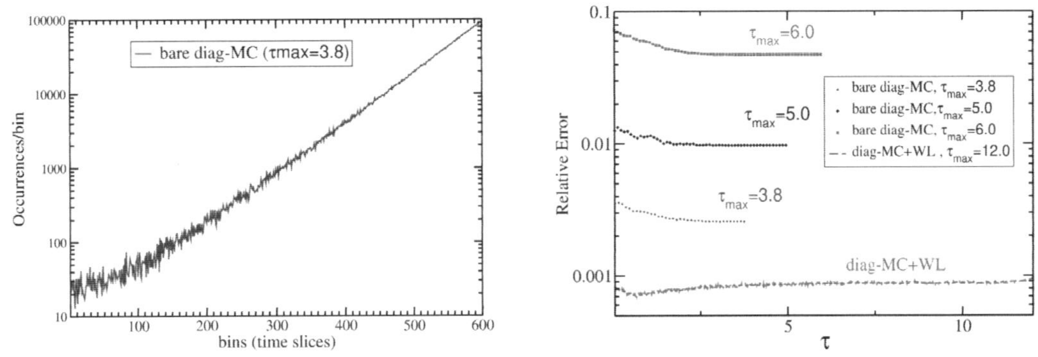

(a) (b)

Figure 4. (a) $G(\tau)$ calculated using bare diag-MC. (b) The relative errors as a function of τ for various values of τ_{max} when we use the bare diag-MC method and when using the Wang-Landau method to make the histogram of $G(\tau)$ flat.

The diag-MC process for fixed τ discussed in Sec. 3 generates a histogram which represents the number of times $N_n^+(\tau)$, and $N_n^-(\tau)$ a given order n appeared in the Markov process with positive or negative sign. Then, we can compute $O_n^{\pm}$ using the ratio of these occurrences, i.e.,

$$O_n^{\pm}(\tau) = \frac{N_n^{\pm}(\tau)}{N_0^{+}(\tau)} O_0^{+}(\tau). \tag{7}$$

Notice that we have used O_0^+ as a reference in both cases because $O_0^- = 0$.

In Fig. 5 we illustrate both contributions $O_n^{\pm}$ obtained by applying the diag-MC+MUCA sampling for 3×10^8 MC steps for $\tau = 4$ and $\tau = 5$. Notice that the two contributions O_n^+ and O_n^- are close and their relative difference decreases exponentially with τ, thus, the statistical error increases exponentially with τ. This is a manifestation of the so-called "sign-problem" in quantum Monte Carlo simulation. In this case the problem with the bare application of the diag-MC discussed in Sec. 3.1, is significantly enhanced due to the increased fluctuations in each of the $N_n^{\pm}$ distributions themselves with increasing τ.

We can apply the flat-histogram technique as discussed in Sec.3.2 to each of the distributions $O_n^{\pm}$ simultaneously as illustrated in Fig. 6(a). This approach by reducing the statistical fluctuations in $O_n^{\pm}(\tau)$, for large τ, allows us to obtain a more accurate $G(\tau)$ as illustrated in Fig. 6(b). In Fig. 6(b) we present $G(\tau)$ for the full t-J model using $J/t = 0.3$ and for $\vec{k} = (\pi/2, \pi/2)$ calculated for 20 different diag-MC runs using diag-MC with (red crosses) and without (open blue circles) the application of the multicanonical approach and for the same number of MC steps in each one of these MC simulations. Notice that as τ increases the statistical fluctuations of $G(\tau)$ obtained with bare diag-MC are larger than those obtained with the application of the flat-histogram technique.

5.2. Sampling the histogram of $G(\tau)$.

We can also apply the flat-histogram approach when we include τ as a dimension of the sampling space for the full t-J model. In this case, as in Sec. 4, we make the histogram of $G(\tau)$ flat.

Once again we wish to demonstrate the need for the application of flat-histogram techniques. In Fig. 7(a) we present the positive and negative contributions $G^{\pm}(\tau)$ to $G(\tau)$ calculated with

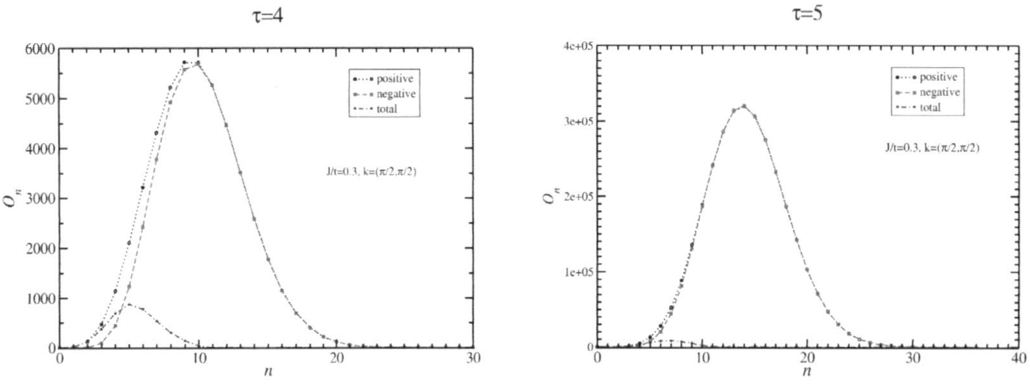

(a) (b)

Figure 5. The calculated $O_n^{\pm}(\tau)$ for the full t-J model using $J/t = 0.3$ and for $\vec{k} = (\pi/2, \pi/2)$ for (a) $\tau = 4$ and (b) $\tau = 5$.

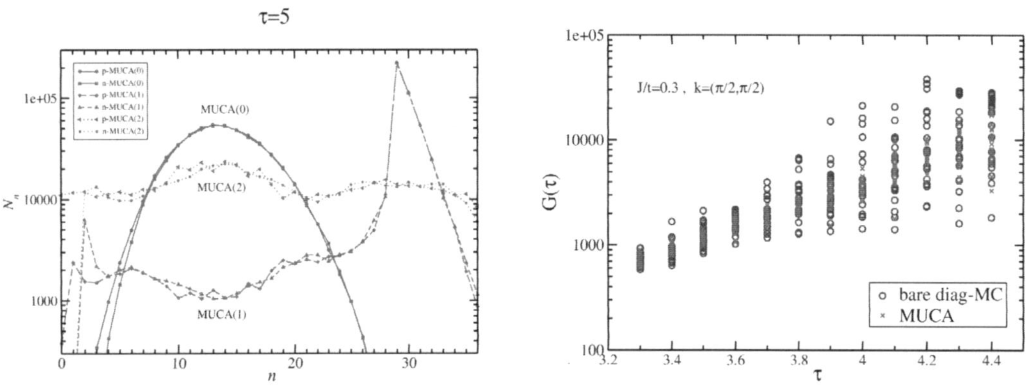

(a) (b)

Figure 6. (a) Demonstration of the application of the flat-histogram approach to make both histograms $N_n^{\pm}(\tau)$ flat as a function of n. This is illustrated for $\tau = 5$ after one application of multicanonical step (MUCA(1)) and after a second multicanonical step (MUCA(2)). (b) $G(\tau)$ calculated for 20 diag-MC runs using diag-MC without (open blue circles) and with (red crosses) the application of flat-histogram technique.

the bare diag-MC and in Fig. 7(b) we present the results of the calculation of $G(\tau)$ and the $G^{\pm}(\tau)$ contributions using diag-MC with the application of the WLA. We note that while both calculations were carried out for approximately the same amount of CPU time (200 repetitions of 10^8 MC steps each), for the bare diag-MC case we used $\tau_{max} = 3.8$, while for diag-MC+WLA we used $\tau_{max} = 4.8$. If we were to extend the value of τ_{max} for the former calculation to the value used in the latter calculation, most like there would be no occurrences of the Markov process in the crucial interval of $\tau = 0$. The reason is that the exponential growth of $G(\tau)$ "pushes" all the occurrences of the Markov process at high value of τ. This is evident for this smaller value of τ_{max} by the large fluctuations which are more clearly noticeable at low values of τ. In Fig. 8(a) we show the relative errors for various values of τ_{max} when calculated with the bare diag-MC method. Notice that the error in $G^{\pm}(\tau)$ for a given number of MC steps grows exponentially with the maximum value of τ_{max} used in this simulation because of the exponential dependence of $G^{\pm}(\tau)$ on τ itself. In Fig. 8(a) we also present the error obtained when the WLA is applied in conjunction with the diag-MC for approximately the same amount of CPU time. This comparison suggests that the gain of the application of the WLA is significant.

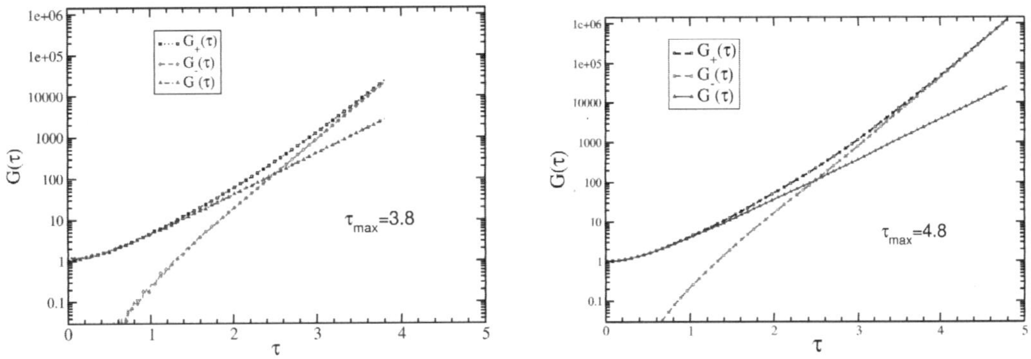

(a) (b)

Figure 7. $G(\tau)$ for the full t-J model (for $J/t = 0.3$ and $\vec{k} = (\pi/2, \pi/2)$) calculated using (a) bare diag-MC and (b) diag-MC+WLA.

In Fig. 8(b) the relative errors of $G^{\pm}(\tau)$ and $G(\tau)$ are shown. We can understand the reduction of error when applying the flat-histogram technique in a very simple way as follows. For a given value of τ the error on both $G^{+}(\tau)$ and $G^{-}(\tau)$ is significantly reduced when combining the diag-MC technique with the flat-histogram technique. The reason for this is essentially the same as in the case of the NCA (Sec. 4) where every diagrammatic contribution is positive definite. As a result the error in the difference $G(\tau) = G^{+}(\tau) - G^{-}(\tau)$ is also significantly reduced. Thus, the flat-histogram technique is reducing the error in $G(\tau)$ itself.

6. Discussion

We demonstrated that the combination of the flat-histogram techniques with the diag-MC method yields a significant improvement over the bare diag-MC method. This combination has been applied to extract the imaginary-time single-hole Green's function $G(\tau)$ in the t-J model.

First, we restricted our sampling space to only those non-crossing diagrams whose contribution is positive-definite and, thus, can be regarded as a probability distribution. We found that the combination of flat-histogram techniques with the diag-MC method yields much more accurate results for $G(\tau)$ over a wide range of the imaginary-time τ.

Second, when we sampled the entire diagrammatic space, without the NCA restriction, there are both positive and negative contributions from non NCA diagrams. The positive and

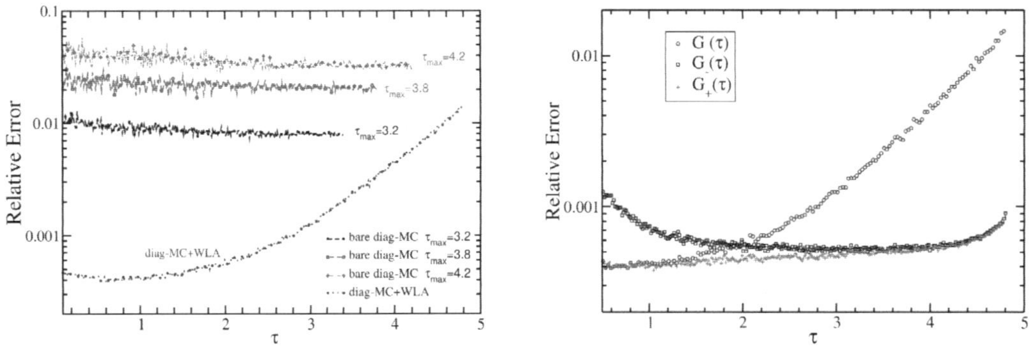

(a) (b)

Figure 8. (a) Relative errors of $G(\tau)$ obtained by using the bare diag-MC for different values of τ_{max} are compared to the error obtained by applying diag-MC+WLA. (b) The relative errors to the quantities shown in part Fig. 7(b).

negative contributions approach each other exponentially as we increase τ, thus, the error in the difference, i.e., $G(\tau)$, grows exponentially with increasing τ. This is very similar to the "sign" problem in other quantum Monte Carlo simulations of fermions where the statistical error increases exponentially as a function of the particle number. The application of the flat-histogram technique with the diag-MC allows us to obtain more accurate results on both the positive and the negative contributions to $G(\tau)$. This enables us to compute $G(\tau)$ in a wider range of τ.

7. Acknowledgements

This work was supported in part by the U.S. National High Magnetic Field Laboratory, which is funded by NSF DMR-1157490 and the State of Florida.

References

[1] D. M. Ceperley, Rev. Mod. Phys. 67, 279 (1995). Rev. Mod. Phys. **71**, S438-443, (1999).
[2] B. A. Berg, and T. Neuhaus, Phys. Lett. B, **267**, 249 (1991).
[3] P. M. C. de Oliviera *et al.*, J Phys. **26**, 677 (1996).
[4] F. Wang and D. P. Landau, Phys. Rev. Lett. **86**, 2050 (2001).
[5] M. Troyer, S. Wessel, and F. Alet, Phys. Rev. Lett. **90**, 120201 (2003).
[6] E. Gull, A. J. Millis, A. I. Lichtenstein, A. N. Rubstov, M. Troyer, P. Werner, Rev. Mod. Phys. **83**, 349 (2011). G. Li, W. Werner, and A. N. Rubstov, S. Base, M. Potthoff, Phys. Rev. B **80**, 195118 (2009). E. Gull, Ph.D. Thesis.
[7] N. G. Diamantis and E. Manousakis, Phys. Rev. E, **88**, 043302 (2013). N. G. Diamantis, and E. Manousakis, Phys. Procedia, **57**, 48 (2014).
[8] H. Fröhlich, H. Pelzer, and S. Zienau, Philos. Mag. **41**, 221 (1950).
[9] N. V. Prokof'ev and B. V. Svistunov, Phys. Rev. Lett. **81**, 2514 (1998).
[10] A. S. Mishchenko, N. V. Prokof'ev, A. Sakamoto, B. V. Svistunov, Phys. Rev. B, **62**, 6317 (2000).
[11] C. L. Kane, P.A. Lee, and N. Read, Phys. Rev. B **39,** 6880 (1989).
[12] E. Manousakis, Rev. Mod. Phys. **61**, 1 (1991).
[13] Z. Liu and E. Manousakis, Phys. B **44,** 2414 (1991).
[14] Z. Liu and E. Manousakis, Phys. B **45,** 2425 (1992).
[15] E. Dagotto, Rev. Mod. Phys. **66**, 763 (1994).

Journal of Physics: Conference Series **750** (2016) 012003 doi:10.1088/1742-6596/750/1/012003

Manifest and Subtle Cyclic Behavior in Nonequilibrium Steady States

R K P Zia[1,2], Jeffrey B Weiss[3], Dibyendu Mandal[3,4] and Baylor Fox-Kemper[5]

Abstract. Many interesting phenomena in nature are described by stochastic processes with irreversible dynamics. To model these phenomena, we focus on a master equation or a Fokker-Planck equation with rates which violate detailed balance. When the system settles in a stationary state, it will be a nonequilibrium steady state (NESS), with time independent probability distribution as well as persistent probability current loops. The observable consequences of the latter are explored. In particular, cyclic behavior of some form must be present: some are prominent and manifest, while others are more obscure and subtle. We present a theoretical framework to analyze such properties, introducing the notion of "probability angular momentum" and its distribution. Using several examples, we illustrate the manifest and subtle categories and how best to distinguish between them. These techniques can be applied to reveal the NESS nature of a wide range of systems in a large variety of areas. We illustrate with one application: variability of ocean heat content in our climate system.

[1]Department of Physics and Astronomy, Iowa State University, Ames, IA 50011, USA
[2]Department of Physics, Virginia Polytechnic Institute and State University, Blacksburg, VA 24061, USA
[3]Department of Atmospheric and Oceanic Sciences, University of Colorado, Boulder, CO 80309, USA
[4]Department of Physics, University of California, Berkeley, CA 94720, USA
[5]Department of Earth, Environmental, and Planetary Sciences, Brown University, Providence, RI 02912, USA

E-mail: rkpzia@vt.edu

1. Introduction

In the standard Boltzmann-Gibbs formulation of equilibrium statistical mechanics, time plays no role. Once $\{q\}$, the set of configurations (or microstates) of the system of interest is chosen, the energy functional (Hamiltonian) $H(q)$ is provided, and the conditions for equilibrium are specified, then $P(q)$, the probability for finding the system in q is known. This structure is built on Boltzmann's fundamental hypothesis: $P(q) \propto \delta(E - H(q))$ for an isolated system with total energy E. From here, various other ensembles and their associated P's follow. The main task is to compute averages of various observable quantities $\mathcal{O}(q)$, $\langle \mathcal{O} \rangle \equiv \Sigma_q \mathcal{O}(q) P(q)$.

The Boltzmann-Gibbs paradigm is clearly inadequate to describe many stochastic processes in nature. In addition to a need to describe time dependent phenomena (e.g., autocorrelations), there are many systems which interact with the environment in a manner that *violates* time reversal. In particular, *all* biological systems consume nutrients and discard waste, with processes that clearly cannot be reversed in time. In this case, the system settles into *non-equilibrium steady states* (NESS), with characteristics absent from systems in thermal

Journal of Physics: Conference Series **750** (2016) 012003 doi:10.1088/1742-6596/750/1/012003

equilibrium. Chief among these are the presence of probability currents and loops [1], much like those in magnetostatics. In this brief note, we report some observable consequences of these current loops, in both manifestly cyclic behavior and more subtle realizations. We will also point to the notion of "probability angular momentum" and relate it to a more familiar quantity: the two point correlation at *unequal* times.

For the simplest stochastic process that leads to NESS, we may start with a master equation for the time dependent $P(q,t)$, with rates that violate detailed balance (DB). In general, this equation takes the form

$$\partial_t P(q,t) = \sum_{q'} \left[W(q' \to q)\, P(q',t) - W(q \to q')\, P(q,t)\right] \equiv \sum_{q'} K(q' \to q), \tag{1}$$

where $W(q' \to q)$ is the transition rate for the system in q' to become one in q and $K(q' \to q)$ is the *net* probability current from q' to q. Now, DB is often displayed as $W(q' \to q)/W(q \to q') = \exp\beta\left[H(q') - H(q)\right]$, so that, for a system in thermal equilibrium with stationary $P^* \propto e^{-\beta H}$, the currents K^* vanish identically. (Note that quantities in the stationary state are associated with *.) By contrast, for processes modeled by W's that violate DB, some K^*'s in the NESS must be non-zero and must form closed loops.

Though the master equation is the most general formulation for this class of stochastic processes, we will restrict ourselves, for simplicity, to configuration spaces described by continuous variables, ξ_α (or just $\vec{\xi}$) in arbitrary dimensions, and evolution controlled by the Fokker-Planck equation (FPE): [1]

$$\partial_t P\left(\vec{\xi},t\right) = \partial^\alpha \left\{\partial^\beta D_{\alpha\beta} P - V_\alpha P\right\} \equiv -\partial^\alpha K_\alpha. \tag{2}$$

Here, D and V represent the diffusive and drift aspects, respectively. A more intuitive description, as well as rules for coding simulations, is the Langevin equation

$$\partial_t \vec{\xi} = \vec{V} + \vec{\eta}, \tag{3}$$

where $\vec{\eta}$ is a Gaussian noise with $\langle\vec{\eta}\rangle = 0$ and $\langle\eta_\alpha(t)\eta_\beta(t')\rangle = D_{\alpha\beta}\delta(t-t')$. As the FPE is just a continuity equation, we can identify the probability current here as $K_\alpha = -\left\{\partial^\beta D_{\alpha\beta}P - V_\alpha P\right\}$. Of course, the stationary distribution, P^*, satisfies $\partial_t P^* = 0$, i.e., $\vec{\nabla}\cdot\vec{K}^* = 0$. In this approach, the dynamics satisfies DB provided $\left[D^{-1}\right]^{\gamma\alpha}\left(\partial^\beta D_{\alpha\beta} - V_\alpha\right)$ is the gradient of some scalar function $s\left(\vec{\xi}\right)$, i.e., $\partial^\gamma s$. Then, it is straightforward to show that $\vec{K}^* \equiv 0$ with $P^* \propto e^{-s}$. Our interest here are processes which violate this condition, when $\vec{K}^*$ is non-trivial. Being divergenceless, it can be expressed (in 3 dimensions) as curl of $\vec{\psi}^*$, the stream function (in the language of fluid dynamics), while $\vec{\nabla}\times\vec{K}^*$ is known as the vorticity, $\vec{\omega}^*$.

2. Mass/fluid *vs.* probability angular momenta; two point correlation functions

Angular momentum, a familiar concept from textbooks on classical mechanics and fluid dynamics, is associated with mass in motion: $\vec{L} = \vec{r}\times m\vec{v}$, $\int d\vec{r}\vec{r}\times\vec{v}\rho(\vec{r})$, and $\int d\vec{r}\left(\vec{r}\times\vec{J}\right)$ (where $\vec{J} = \rho\vec{v}$ is the fluid current). We transfer this fluids concept to probability by considering the mapping $\{\vec{r},\rho,\vec{J}\} \to \{\vec{\xi},P,\vec{K}\}$, so that $\int d\xi\left(\vec{\xi}\times\vec{K}\right)$ is a quantity of interest and naturally named "probability angular momentum." Of course, in arbitrary dimensions, rotations and

[1] The Einstein summation convention is used here.

angular momenta are not vectors, but pseudo tensors. Thus, instead of a vector $\vec{L}$, we should consider a matrix $\mathbb{L}$, with elements $L_{\alpha\beta} = -L_{\beta\alpha}$ and

$$L_{\alpha\beta} = \int \mathrm{d}\vec{\xi} \left(\xi_\alpha K_\beta - \xi_\beta K_\alpha\right). \tag{4}$$

Note that, due to the normalization condition $\int \mathrm{d}\vec{\xi}\, P = 1$, our "mass" is unity, so that the unit of $\mathbb{L}$ is just ξ^2/t, precisely that of diffusion. This remarkable feature is not coincidental, as the intimate connection between them will be shown in the next section. We further note that, in a NESS, we have $\int \mathrm{d}\vec{\xi}\, \vec{K}^* = 0$, so that $\mathbb{L}^*$ is independent of the choice of the origin of $\vec{\xi}$ (and is just the total vorticity). Other familiar concepts such as angular velocity $\vec{\Omega}$ and inertia tensor, $\mathbb{I}$, also have analogs. Specifically, we see that $\int r_i r_j \rho$ maps to the two point correlation $\langle \xi_\alpha \xi_\beta \rangle \equiv \int \xi_\alpha \xi_\beta P$. Meanwhile, in analog with $\vec{v} = \vec{\Omega} \times \vec{r}$, we can define our angular velocity via $K_\alpha = \Omega_\alpha^\beta \xi_\beta P$ [2] and come up with the equivalent of $\vec{L} = \mathbb{I}\vec{\Omega}$.

We end this section with a not-so-familiar generalization of angular momentum. If the trajectory of a point mass $\vec{r}(t)$ is known, then we may consider the quantity $\vec{a} \equiv \vec{r}(t) \times \vec{r}(t')$, the magnitude of which is just the area of the parallelogram spanned by the two vectors. Clearly, it is related to the angular momentum by $\vec{L} = m\partial_{t'}\vec{a}|_{t'=t}$. The analogous generalization to $L_{\alpha\beta}$ here is a special two point correlation function:

$$\tilde{C}_{\alpha\beta}(t,t') \equiv \int \mathrm{d}\vec{\xi} \int \mathrm{d}\vec{\xi}'\, \mathcal{P}\left(\vec{\xi},t;\vec{\xi}',t'\right) \left[\xi_\alpha \xi'_\beta - \xi_\beta \xi'_\alpha\right], \tag{5}$$

where $\mathcal{P}$ is the *joint* probability. If $t' > t$, then $\mathcal{P}$ is the product of the conditional probability, $\mathcal{G}\left(\vec{\xi}',t'|\vec{\xi},t\right)$, and $P\left(\vec{\xi},t\right)$. Regarding the FPE (Eq. 2) as a Schrödinger equation, $\mathcal{G}$ is the familiar propagator in quantum mechanics. As in the point mass case, we have

$$L_{\alpha\beta} = \partial_{t'} \tilde{C}_{\alpha\beta}\Big|_{t'=t}. \tag{6}$$

Meanwhile, we see that $\tilde{C}_{\alpha\beta}(t,t')$ is just twice the antisymmetric part of the two point correlation at arbitrary times, $\langle \xi_\alpha \xi_\beta \rangle_{tt'}$. Being antisymmetric, it is necessarily odd under exchange $t \Leftrightarrow t'$. In the stationary state, translation invariance prevails and so, $\tilde{\mathbb{C}}^*$ depends only on the difference $\tau \equiv t' - t$. Since it is *odd* under time reversal, $\tilde{\mathbb{C}}^* \neq 0$ is a concrete measure of DB violation and irreversibility in a stationary state.

While the framework presented above is valid for all stochastic processes, it is valuable to illustrate these ideas in an explicitly solvable system: the Linear Gaussian Model (LGM) [3, 2, 1]. In an LGM, $D_{\alpha\beta}$ (elements of $\mathbb{D}$) are constants, while $\vec{V}$ is linear in $\vec{\xi}$, as in generalized simple harmonic oscillators (SHO): $\vec{V} = \mathbb{A}\vec{\xi}$. (Refer to Eqs. 2 and 3.) Thus, the model is completely defined by two matrices, $\mathbb{D}$ and $\mathbb{A}$. Of course, $\mathbb{D}$ must be positive symmetric, while the real parts of the eigenvalues of $\mathbb{A}$ must be negative (for the stability of the process). If $\mathbb{D}^{-1}\mathbb{A}$ is symmetric, then DB is satisfied and the scalar s is $-\vec{\xi}\cdot\mathbb{D}^{-1}\mathbb{A}\vec{\xi}/2$. If not, then the stationary P^* is still a Gaussian[3]: $P^* \propto \exp\left\{-\vec{\xi}\cdot\mathbb{C}^{-1}\vec{\xi}/2\right\}$, where $\mathbb{C}$ is fixed by [2] $\mathcal{S}[\mathbb{AC}] = -\mathbb{D}$ and $\mathcal{S}$ stands for "the symmetric part of." Clearly, $\mathbb{C}$ is the covariance matrix in the steady state, i.e., the equal time two point function. (Note that $\mathbb{C}$ is not the same as $\tilde{\mathbb{C}}$!) Meanwhile, we have $\vec{\xi}P^* = -\mathbb{C}\vec{\nabla}P^*$, which leads to $\vec{K}^* = -\left\{\mathbb{D}\vec{\nabla} - \mathbb{A}\vec{\xi}\right\}P^* = -[\mathbb{AC}+\mathbb{D}]\vec{\nabla}P^*$. Since $\mathbb{D} = -\mathcal{S}[\mathbb{AC}]$, the sum $[\mathbb{AC}+\mathbb{D}]$ is $\mathcal{A}[\mathbb{AC}]$, the *antisymmetric* part of $\mathbb{AC}$. Thus, $\vec{K}^*$ is manifestly divergence free while the stream function (a matrix here) can be identified as $-[\mathbb{AC}+\mathbb{D}]P^*$. Moreover, it is straightforward to obtain an explicit expression for $\mathbb{L}^*$:

$$\mathbb{L}^* = -2\mathcal{A}[\mathbb{AC}] \tag{7}$$

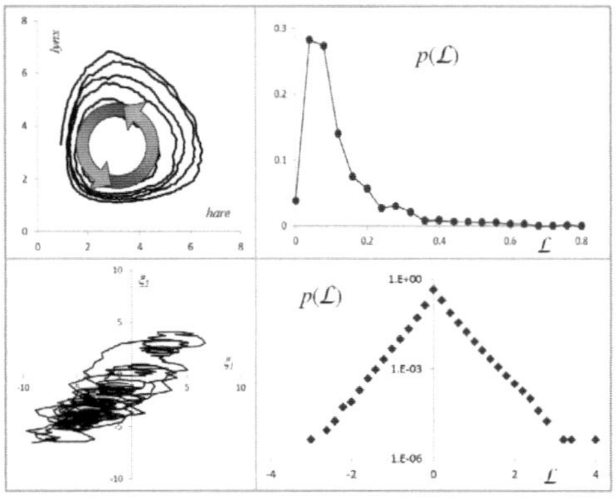

Figure 1. Short trajectories and distributions $p(\mathcal{L})$ for two simple stochastic processes: noisy Lotka-Volterra (upper panels) and SHO's coupled to thermal baths at different temperatures (lower panels). In the former, cyclic behavior is manifest, associated with a "one-sided" $p(\mathcal{L})$. By contrast, this behavior is quite subtle in the latter system, with a distribution that is almost symmetric (around $\mathcal{L} = 0$).

and its generalization:

$$\tilde{\mathbb{C}}^*(\tau) = -2\mathcal{A}\left[e^{\mathbb{A}\tau}\mathbb{C}\right] \tag{8}$$

In this setting, we see that diffusion ($\mathbb{D}$) and angular momentum ($\mathbb{L}^*/2$) are just the symmetric and antisymmetric parts of one matrix: $-\mathbb{AC}$. Thus, they must have the same units, and play complementary roles in any stochastic process. Finally, note that the angular velocity matrix is given by $\Omega^* = \mathbb{D}\mathbb{C}^{-1} + \mathbb{A}$ [2], while $\mathbb{L}^* = 2\mathbb{C}\Omega^*$ is the analog of $\vec{L} = \mathbb{I}\vec{\Omega}$.

3. Distributions of $\mathbb{L}$

Having established that there must be some cyclic behavior in any NESS, we ask if this feature is displayed (a) prominently and manifestly, or (b) in some obscure and subtle way. To answer which category a system belongs to, we must go beyond the average values of angular momenta (Eqs. 4 and 7) and study the full distribution: $p(L_{\alpha\beta}) \equiv \int \delta\left(L_{\alpha\beta} - [\xi_\alpha K_\beta - \xi_\beta K_\alpha]\right)$. For simplicity, let us consider a two-dimensional $\vec{\xi}$-space, so that there is only one independent component in any antisymmetric matrix, e.g., $\langle\mathcal{L}\rangle \equiv L^*_{12}$. Then, our task simplifies to the study of $p(\mathcal{L})$. From simulations, it can be obtained by computing $\mathcal{L}(t)$ from a long trajectory $\vec{\xi}(t)$ (in the steady state), and compiling a histogram.

Clearly, the most extreme example in category (a) is a deterministic orbit (e.g., Keplerian) which yields a fixed angular momentum: $p = \delta(\mathcal{L} - const)$. More common, stochastic systems of this type will display broader p's. To illustrate, we consider a stochastic Lotka-Volterra model [4] for the population of hares and lynx, ξ_h and ξ_l respectively. A specific example, [2] associated with the upper panels of Fig. 1, is:

$$\dot{\xi}_h = \xi_h[2 - \xi_l], \quad \dot{\xi}_l = \xi_l[-4 + \xi_h] \tag{9}$$

plus noise. The figure shows a typical trajectory in the space of hare/lynx populations from a simple simulation run, as well as the associated distribution $p(\mathcal{L})$. Clearly, the latter is the

[2] We caution that this model is designed to illustrate properties in category (a) systems. It is too simplistic to provide a good description of the full complexity of predator-prey behavior. See [5] for a good treatment. For a recent review on stochastic LV systems, see, e.g., [6].

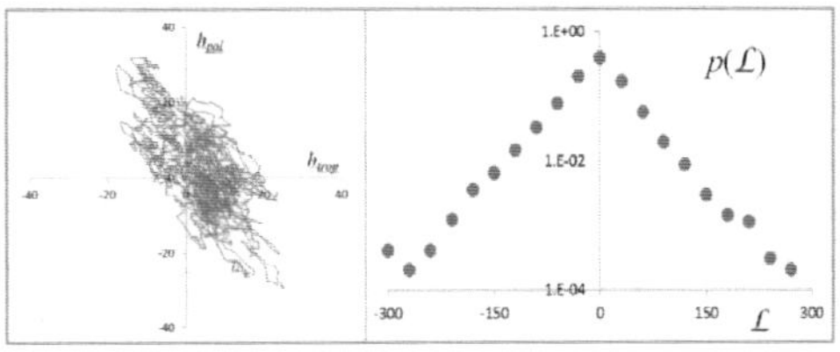

Figure 2. A short trajectory and distribution $p(\mathcal{L})$ for the ocean heat content anomalies found in the tropical and polar regions (in a millennium long run with the Community Earth System Model). The units of h is $ZJ = 10^{21}$ Joules. The units for $\mathcal{L}$ is ZJ^2 per season. The distribution indicates that cyclic behavior in these two variables is very subtle.

result of noise on a δ distribution. Since the dynamics of such models manifestly violate time reversal symmetry and DB, a distribution dominated by one sign of $\mathcal{L}$ is expected. Turning to the subtler NESS in category (b), the average $\langle\mathcal{L}\rangle$ can be quite small, the result of a broad $p(\mathcal{L})$ with only a slight asymmetry in $\mathcal{L}$. Let us illustrate with two SHO's coupled to thermal baths at different T's [7]. Starting with the standard Hamiltonian $\mathcal{H} = \left[k_1\xi_1^2 + k_2\xi_2^2 + k_\times(\xi_1 - \xi_2)^2\right]/2$, we model the effects of the thermal baths by

$$\dot{\xi}_\alpha = -\lambda_\alpha\left(\partial\mathcal{H}/\partial\xi_\alpha\right) + \eta_\alpha, \tag{10}$$

with $\langle\eta\rangle = 0$ and $\langle\eta_\alpha(t)\eta_\beta(t')\rangle = 2\lambda_\alpha T_\alpha \delta_{\alpha\beta}\delta(t-t')$. Since this system is precisely an LGM, we find $\langle\mathcal{L}\rangle \propto k_\times(T_1 - T_2)$, which shows that it settles into a NESS only when $k_\times \neq 0$ and $T_1 \neq T_2$. In the lower left panel of Fig. 1, we show a typical trajectory, which displays no obvious preferential rotation. The right panel shows a nearly symmetric $p(\mathcal{L})$, with barely discernible asymmetry. As a result, the average $\langle\mathcal{L}\rangle \sim 4.70 \times 10^{-3}$ is much smaller than the standard deviation $\Delta\mathcal{L} \sim 32.41 \times 10^{-3}$. Here, we find $\langle\mathcal{L}\rangle$ to be entirely consistent with the theoretical prediction of 4.66×10^{-3}. Finally, we note that the large $\mathcal{L}$ behaviors are exponential, governed by different decay constants. These can also be computed, since we can obtain analytically the full $p(\mathcal{L})$ of Gaussian models (details to be published elsewhere [8]).

Apart from these simple examples, we have recently studied other systems in NESS, including an epidemic model with asymmetric infection rates [9] and a heterogeneous non-linear q-voter model [10]. Both settle into NESS that display only very subtle cyclic behavior. Here, let us present preliminary results concerning a much larger and complex system: variations in the heat content of our oceans. As a stochastic process, the oceans are heated in the tropical regions and suffer loss mostly from the polar regions, forming clearly a non-equilibrium system. Over long periods, it appears quasi-stationary and may be regarded as a NESS. Unfortunately, high quality data for these anomalies in the real oceans form only a small set and date from about half a century ago [11]. Nevertheless, we can combine the data into two time series, $h_{\text{tropics}}(t)$ and $h_{\text{polar}}(t)$, and study both $\langle\mathcal{L}\rangle$ and $p(\mathcal{L})$. The results are very similar to those of the SHO's in Fig. 1. The details are quite complicated and will be published elsewhere [8]. Here, we turn to a much longer (about a millennium) data set, created using the state-of-the-art Community Earth System Model [12]. Though non-trivial complications concerning the analysis also exist here [8], the results are consistent with the those from real data. Illustrating with a small portion of this trajectory and showing the $p(\mathcal{L})$ in Fig. 2, we recognize that these NESS aspects are similar to those in the two-temperature SHO case. However, there is a subtle difference: $\langle\mathcal{L}\rangle \sim -5.4\, ZJ^2/season$ is negative, a sign naively opposite to that in the SHO's. This difference indicates that temperature difference is not the only controlling factor for the sign of $\langle\mathcal{L}\rangle$. The more significant factor is the level of the noise associated with each degree of freedom. Details of this kind of analysis and understanding will be presented elsewhere [8].

4. Summary and Outlook

We have shown that the angular momenta $\mathbb{L}^*$ and the two point correlation at unequal times $\tilde{\mathbb{C}}^*(\tau)$ are, for *any* NESS, excellent measures of the underlying time-reversal and DB violating dynamics. Beyond these simplest quantities, any multipoint correlation functions at unequal times will provide a platform for measuring such characteristics.

We have also shown that, for certain systems in NESS, DB violation is so patently obvious that the irreversible nature is prominent and manifest. For example, hares do *not* prey on lynxes! On the other hand, many NESS do not display such clear behavior, as the time-reversal violating aspects are more obscure and subtle. Clearly, it is desirable to understand more deeply the mechanisms which control the outcomes of a system. What are the parameters which, when varied continuously, will take a system from category (a) to (b)? Is the "transition" abrupt and discontinuous? or smooth and continuous? Is it possible that $p(\mathcal{L})$ displays a combination of both "components"? We believe this is a promising and rich avenue for future research, both as a novel measure to characterize different systems in NESS and as a possible step towards a overarching framework for the foundations of non-equilibrium statistical mechanics.

5. Acknowledgments

We are grateful to A Mellor, M Mobilia, A D Nelson, B Schmittmann and U C Täuber for enlightening discussions. This research is supported partly by a US National Science Foundation grant: OCE-1245944.

6. References

[1] Zia R K P and Schmittmann B 2007 *J. Stat. Mech. Theory Exp.* **2007** P07012
[2] Weiss J B 2007 *Phys. Rev.* **E76** 061128
[3] Lax M 1966 *Rev. Mod. Phys.* **38** 541
[4] Lotka A J 1920 *Proc. Natl. Acad. Sci. U.S.A.* **6** 410; Volterra V 1926 *Mem. Acad. Lincei Roma* **2** 31
[5] McKane A J and Newman T J 2005 *Phys. Rev. Lett.* **94** 218102
[6] Mobilia M, Georgiev I T and Täuber U C 2007 *J. Stat. Phys.* **128** 447
[7] Täuber U C and Zia R K P 1998 unpublished
[8] Nelson A D, Mandal D, Fox-Kemper B, Weiss J B, and Zia R K P 2016 to be published
[9] Shkarayev M S and Zia R K P 2014 *Phys. Rev.* **E90** 032107
[10] Mellor A, Mobilia M, and Zia R K P 2016 EPL, in press
[11] https://www.nodc.noaa.gov/OC5/3M_HEAT_CONTENT/basin_data.html
[12] http://www2.cesm.ucar.edu/

Journal of Physics: Conference Series **750** (2016) 012004 doi:10.1088/1742-6596/750/1/012004

Thermodynamics and phase coexistence in nonequilibrium steady states

Ronald Dickman

Departamento de Física and National Institute of Science and Technology for Complex Systems,
ICEx, Universidade Federal de Minas Gerais,
C. P. 702, 30123-970 Belo Horizonte, Minas Gerais - Brazil

E-mail: dickman@fisica.ufmg.br

Abstract. I review recent work focussing on whether thermodynamics can be extended to nonequilibrium steady states (NESS), in particular, the possibility of consistent definitions of temperature T and chemical potential μ for NESS. The testing-grounds are simple lattice models with stochastic dynamics. Each model includes a drive that maintains the system far from equilibrium, provoking particle and/or energy flows; for zero drive the system relaxes to equilibrium. Analysis and numerical simulation show that for spatially uniform NESS, consistent definitions of T and μ are possible via coexistence with an appropriate reservoir, if (and in general only if) a particular kind of rate (that proposed by Sasa and Tasaki) is used for exchanges of particles and energy between systems. The program fails, however, for nonuniform systems. The functions T and μ describing isolated phases cannot be used to predict the properties of coexisting phases in a single, phase-separated system.

1. Introduction

In recent decades the possibility of extending thermodynamics to far-from-equilibrium steady states has attracted considerable attention. By "thermodynamics" one understands a macroscopic description employing a small number of variables [1], although the set of variables needed to describe a nonequilibrium system may be somewhat larger than required for equilibrium. Thus in near-equilibrium thermodynamics, fluxes of mass, energy, and other conserved quantities enter the description [2,3]. This approach functions well in the hydrodynamic or local-equilibrium regime; here, by contrast, we are concerned with systems maintained far from equilibrium, in which the stationary probability distribution is very different from the Boltzmann distribution. Oono and Paniconi argued for the possibility of a steady-state thermodynamics (SST) on general grounds [4]. Sasa and Tasaki [5], extending their ideas, proposed definitions of the chemical potential and pressure in SST, and developed a theoretical analysis of the driven lattice gas; a numerical implementation in driven systems is discussed in [6]. More recently, Pradhan et al. [7] tested, via numerical simulation of driven lattice gases in contact, the consistency of the scheme proposed in [5], focussing on the validity of the zeroth law. They found that the zeroth law holds to good approximation, and suggested that observed deviations are due to nonuniformities induced by the contact itself. Of course, thermodynamics should hold *exactly* (not merely to good approximation), so it is essential to understand if the deviations are the result of approximations or are intrinsic to the theory.

Journal of Physics: Conference Series **750** (2016) 012004 doi:10.1088/1742-6596/750/1/012004

While equilibrium thermodynamics has attained a state of maturity permitting an axiomatic formulation in terms of an all-inclusive entropy function [1], SST is far from such a stage. Rather than proposing formal schemes, I prefer to examine whether key intensive variables, such as chemical potential and temperature, can be defined operationally via *coexistence*. I proceed through a series of increasingly complex examples, beginning with spatially uniform systems before turning to nonuniform situations. I adopt, as minimal criteria for judging the validity of SST, that it be consistent, in the sense of the Zeroth Law, and that it have predictive value. For example, knowledge of the temperature and/or chemical potential of uniform systems as a function of the control parameters should permit one to predict the macroscopic properties of nonuniform systems.

I study two-dimensional stochastic lattice gases which are well established in equilibrium as simple models of fluids, and which, when perturbed from equilibrium, have a plausible interpretation as a fluid subject to an external drive or nonuniform temperature. Most of the conclusions are based on dynamic Monte Carlo simulations. Since the studies reported here are done far from any critical points, finite-size effects are minimal, as confirmed by comparing results for systems of linear size L ranging from 100 to 400 or 800. In the following section I discuss background and basic definitions. Section 3 is devoted to spatially uniform NESS, while in section 4 I turn my attention to nonuniform systems. Section 5 contains a brief summary and conclusions.

2. Intensive parameters, coexistence and reservoirs

Following the ideas of [4-6], I adopt *coexistence* as the basis for defining intensive parameters. Consider two systems, A and B, in steady states, that are permitted to exchange particles and/or energy. If the net flow of these quantities is zero, we say that A and B coexist with respect to (wrt) such exchanges, and assert that they share the same value of the associated intensive parameter. (If both A and B are each at equilibrium, "coexistence" implies *equilibrium* wrt exchange of the quantity of interest.) Thus if A and B coexist wrt both energy and particle exchanges, $T_A = T_B$ and $\mu_A = \mu_B$. (Here I consider single-component systems; the generalization to multicomponent systems is direct. In principle one can also define pressure in an analogous manner, but this is not required in the present discussion.) The Zeroth Law of thermodynamics states that if A and B coexist (wrt a certain type of exchange), and B and C coexist, then A and C must also coexist.

Now if one system, say B, is a *reservoir* of known temperature and/or chemical potential, we can use it to measure the corresponding properties of system A. The reservoir R is defined as having the following characteristics:

> Its intensive properties are invariant under exchange of particles and/or energy;
>
> Its mode of interaction with another system depends only on the intensive properties of the reservoir.

Suppose, for example, that R is a particle reservoir, characterized only by its dimensionless chemical potential $\mu^* = \mu/T$. (Here and in what follows I set Boltzmann's constant $k_B = 1$.) Let system A be an athermal lattice gas (i.e., in which the only interactions are of excluded volume) in which sites can be occupied by at most one particle. Then a suitable scheme of exchange between R and A is as follows. At rate w_R, each particle in A moves to the reservoir, and at rate w_I, each open site in A receives a particle from R. (An open site is one that may be occupied without violating the excluded-volume constraints.) A *could be* a simple lattice gas (no interactions beyond on-site exclusion), in equilibrium, in which case the mean fraction of occupied sites ρ follows $\dot{\rho} = -w_R\rho + w_I(1-\rho)$, while equilibrium statistical mechanics implies that $\rho_{eq} = z/(1+z)$ with $z = \exp[\mu^*]$. Equating $\dot{\rho}$ to zero at equilibrium, we have $w_I/w_R = \exp[\mu^*]$. Since w_R and w_I are properties of the reservoir not the system with which it makes contact, this relation holds for exchange with *any* athermal lattice gas.

We use this simple observation to assert that, if A (an athermal lattice gas, in or out of equilibrium) is in contact with a particle reservoir R, and the net flux of particles between A and R is zero, then the dimensionless chemical potential of A is that of R: $\mu_A^* = \mu_R^*$. This definition of μ^* is equivalent to that proposed by Sasa and Tasaki [5]. For lattice gases with interactions beyond those of excluded volume (for example the case of nearest-neighbor attractive interactions discussed below), particle transfers in general imply energy transfers, and R should be a heat and particle reservoir [8].

3. Uniform driven systems

The considerations of the previous section lead directly to a definition of the chemical potential in athermal lattice gases. Let ρ_{op} be the fraction of *open* sites, that is, sites at which a particle may be inserted without violating the excluded-volume constraints. Exchange between a system A and a particle reservoir R leads to $\dot{\rho} = -w_R\rho + w_I\rho_{op}$, so that at coexistence,

$$\mu^* = \ln \frac{w_I}{w_R} = \ln \frac{\rho}{\rho_{op}} \tag{1}$$

This is convenient computationally: in a simulation at fixed ρ, we simply monitor the fraction of open sites in the stationary state to estimate ρ_{op}; explicit exchange with a reservoir is not needed.

Now suppose that A and B are both athermal lattice gases, in spatially uniform steady states under their respective stochastic dynamics. Particle exchange between A and B may be implemented as follows: a fraction p_r of all transitions are exchange attempts, in which sites $\mathbf{x}$ and $\mathbf{y}$ are chosen at random in systems A and B, respectively. If $\mathbf{x}$ is occupied and $\mathbf{y}$ open, the particle at $\mathbf{x}$ is transferred to $\mathbf{y}$, and vice-versa. Now, in the limit $p_r \to 0$, each transfer is followed by an infinite sequence of transitions *within* systems A and B, allowing them to attain statistical independence. This permits us to write,

$$\dot{\rho}_A = -p_r[\rho_A\rho_{op,B} + \rho_B\rho_{op,A}]. \tag{2}$$

Thus the zero-flux condition (coexistence) implies $\mu_A^* = \mu_B^*$, where the chemical potentials are those determined via contact with a reservoir. Moreover, if $\mu_A^* > \mu_B^*$, Eq. (2) implies a net flux of particles from A to B, and vice-versa, so that the final steady state under particle exchange is predicted by equating the (dimensionless) chemical potentials. (Here it is important to note that ρ_{op} is a nonincreasing function of density ρ, making μ^* an increasing function of density, as required by thermodynamic stability.)

The exchange scheme defined above is called *weak global exchange*: weak in the sense that $p_r \to 0$, global in the sense that exchange is allowed between any pair of sites. In the case of spatially uniform athermal systems under weak global exchange, the chemical potential, Eq. (1), satisfies the Zeroth Law and is predictive of the final state. In fact, the global exchange condition can be relaxed: even if $\mathbf{x}$ and $\mathbf{y}$ are *fixed sites* (representing the more realistic situation of local contact between two systems), the limit $p_r \to 0$ guarantees that the two systems can relax to spatially uniform distributions between exchange events [9].

I note briefly the generalization of the above scheme to continuous-space models. Let A be an athermal system and R a particle reservoir at chemical potential μ^*. Each particle in A has a rate ω to move to R, while R attempts to insert particles in A, at randomly chosen positions, at rate $\omega z V_A$, where V_A is the volume of A. Attempted insertions that satisfy all excluded-volume constraints are accepted. Denoting the acceptance probability for insertion as p_{ins}, we have for the mean number of particles in A, $d\langle N_A\rangle/dt = -\omega[\langle N_A\rangle - zp_{ins}V_A]$, or $\dot{\rho_A} \propto -\rho_A + zp_{ins}$, so that at coexistence, $\mu^* = \ln[\rho_A/p_{ins}]$. (Here ω is an arbitrary rate; weak exchange corresponds to the limit $\omega \to 0$.) In equilibrium, this formula corresponds to the Widom insertion method

Journal of Physics: Conference Series **750** (2016) 012004 doi:10.1088/1742-6596/750/1/012004

[10]. For exchange between two athermal systems A and B, we assume transfer attempts from A to B at rate $\omega\rho_A$ and vice-versa. The zero-flux condition again corresponds to $\mu_A^* = \mu_B^*$.

The proposal outlined above for SST of uniform athermal systems has been tested on the lattice gas with nearest-neighbor exclusion (NNE) [8]. In this model the pair interaction is infinite if the distance between particles is ≤ 1, and is zero otherwise. The equilibrium NNE lattice gas has been studied as a discrete-space version of the hard-sphere fluid, and is known to exhibit a continuous (Isinglike) phase transition to sublattice ordering at a density of $\rho_c \simeq 0.36774$ [11].

Consider a NNE system on a lattice of L^d sites, with a fixed number of particles, and with a stochastic dynamics (e.g., particle hopping). Any particle displacement satisfying the NNE condition is accepted. If the transition rates obey detailed balance, the system relaxes to equilibrium.

The NNE lattice gas exhibits NESS under a drive that biases hopping [12], provided that boundaries along the drive direction are periodic. For single-particle nearest-neighbor hopping on the square lattice, we parameterize the drive such that hopping attempts in the $+x$ direction occur with probability $p/2$, and those in the opposite direction with probability $(1-p)/2$; the attempt probabilities for hopping in the $\pm y$ directions are $1/4$ as in equilibrium. Thus $p = 1/2$ corresponds to the unbiased (equilibrium) case, and $p = 1$ to maximum drive.

Simulating the NNE system on periodic lattices, I determine $\mu^*(\rho, p)$ for $p = 1/2$ (equilibrium) and $p = 1$ (maximum drive); the results are shown in Fig. 1. In the same figure, pairs of adjacent points represent the densities of driven and undriven systems under weak global exchange; the coexisting densities correspond, in fact, to equal values of μ^*, verifying the zeroth law, and showing that we can predict the densities of a driven system and nondriven system at coexistence by equating their respective chemical potentials.

An analogous scheme, for defining the chemical potential μ *and* temperature T of a nonathermal system in a NESS, is developed in [8]; the simplest example is the driven lattice gas with nearest-neighbor interactions, known as the Katz-Lebowitz-Spohn (KLS) model [13]. The interaction energy is $E = -\sum_(i,j\rangle \sigma_i \sigma_j$ where $\sigma_i = 0(1)$ denotes site i vacant (occupied), and the sum is over NN pairs of sites. (Each site may harbor at most one particle, but there is no prohibition of NN occupancy, as there is in the NNE model discussed above.) In equilibrium, this system is equivalent to the Ising model; on the square lattice the critical temperature is $T_c = 0.5673...$. The system evolves via particle-conserving NN hopping dynamics with a drive $\mathbf{D} = D\mathbf{i}$ favoring particle displacements along the $+x$ direction and inhibiting those in the opposite sense. (As before, we assume periodic boundaries along the drive.) The acceptance probability for a particle displacement $\Delta\mathbf{x}$ (to an unoccupied neighbor) is $p_a = f[(\Delta E - \mathbf{D}\cdot\Delta\mathbf{x})/T_{\mathcal{R}}]$, where the function $f(x)$ satisfies detailed balance, i.e., $f(-x) = e^x f(x)$, so that the system can achieve equilibrium if the drive $D = 0$. To begin we consider the widely used Metropolis acceptance probability,

$$p_a = \min\{1, \exp[-(\Delta E - \mathbf{D}\cdot\Delta\mathbf{x})/T_{\mathcal{R}}]\}. \tag{3}$$

Successful implementation of SST requires that we recognize two crucial points. First, on average, any nonzero drive performs work on the system (it facilitates transitions with $\Delta E > 0$ more than those with $\Delta E < 0$). To maintain a steady state, energy must flow to a heat reservoir $\mathcal{R}$, which is intrinsic to the KLS setup. In the above expressions for p_a, $T_{\mathcal{R}}$ is the temperature of this reservoir; since energy flows steadily from particle system to the reservoir, we might expect the temperature T_{KLS} of the particle system to be *greater than* $T_{\mathcal{R}}$.

Second, to measure the temperature and chemical potential, T_{KLS} and μ_{KLS}, we place the particle system in contact with a heat and particle reservoir R (distinct, of course, from $\mathcal{R}$), with parameters T and μ, and search for values that render the fluxes of particles *and* energy zero. Naively, one might suppose that exchanges of particles with reservoir R could be implemented using Metropolis rates, that is,

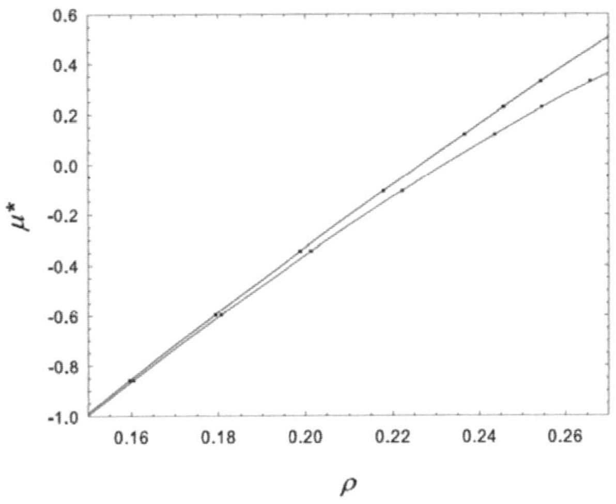

Figure 1. NNE lattice gas: simulation results for μ^* in equilibrium (black curve) and under maximum drive (blue curve), system size $L = 80$. Pairs of points sharing the same value of μ^* represent coexisting densities in the equilibrium and driven systems under weak exchange. Uncertainties are smaller than line thickness and symbol size.

$$w_{ins} = \omega \min\{1, e^{(\mu+j)/T_R}\} \tag{4}$$

and

$$w_{rem} = \omega \min\{1, e^{-(\mu+j)/T_R}\} \tag{5}$$

with j the number of occupied NNs of the target site, on insertion, and the number of occupied NNs of the particle to be removed, in the second case. As shown analytically in [8], *such rates lead to violations of the Zeroth Law.* The only exchange rates that yield consistent definitions of temperature and chemical potential are Sasa-Tasaki (ST) rates, which, in the present instance, follow,

$$w_{ins} = \omega \tag{6}$$

and

$$w_{rem} = \omega e^{-(\mu+j)/T_R} \tag{7}$$

The physical interpretation of ST rates is that, to be transferred, a particle must overcome an energy barrier; the energy of the state following the transfer has no influence on the rate of transfer.

Summarizing, for uniform systems it is possible, subject to certain restrictions, to define consistently intensive parameters such as temperature and chemical potential; equating these parameters enables one to predict the properties of coexisting systems in NESS.

4. Nonuniform systems

Given the encouraging results for uniform systems, I turn to nonuniform NESS. The macroscopic behavior of systems subject to a nonuniform drive, or in which nonuniformities arise due to phase separation, are important potential applications of SST.

To begin I consider the NNE lattice gas subject to a nonuniform drive [9]. I study a lattice of $L \times L$ sites (with L ranging from 200 to 800) divided into two parts by applying a drive in half of the system. The boundaries between the driven and undriven regions are parallel to the drive. Of particular interest are the stationary density and chemical potential profiles ($\rho(j)$ and $\mu^*(j)$, respectively), in the direction perpendicular to the drive. If SST functions correctly, the coexisting bulk densities ρ_0 and ρ_D should be given by the equal chemical potential condition (implying $\rho_D > \rho_0$), and the chemical potential profile should be flat, i.e., $\mu^*(j) = \mu^*(\rho_0, 0) = \mu^*(\rho_D, D)$.

Initially, particles are distributed uniformly over the lattice. For the densities and system sizes considered here, the system attains a stationary state well before 5×10^6 time units, the time allowed for relaxation; averages are calculated over a period of 5×10^6-10^7 time units, following relaxation.

As shown in Fig. 2, the stationary state of the half-driven system is very different from that predicted by SST. The density is higher in the *undriven* region, contrary to the prediction obtained equating the chemical potentials of driven and undriven systems. The chemical potential profile is clearly nonuniform, more so than if the particle density were uniform at the average global density. Particles have migrated so as to increase severalfold the difference in chemical potential between the two regions, rather than diminish it! Notice that this discrepancy exists regardless of how, or whether, we define the chemical potential: Densities which coexist in uniform systems do not coexist when the systems make contact along an edge. (One might suppose that reducing the rate of particle hopping across the boundary between driven and undriven regions might resolve the conflict, as this would correspond to weak exchange between the regions. Such a reduction in fact only serves to accentuate the discrepancy [9].)

Finally, I consider phase coexistence in lattice gases with attractive interactions. Here nonuniformities arise spontaneously, due to phase separation, rather than being imposed via a nonuniform drive. The KLS model provides a simple realization of phase separation far from equilibrium. At high temperatures, the state with uniform density is stable, but (in two or more dimensions) below a certain critical value of T_R, the system segregates into high- and low-density phases separated by a narrow interface. In simulations, the coexisting densities ρ_L and ρ_V are determined by analyzing the density profiles in the direction perpendicular to the drive [14]. In these studies ST rates are used for hopping in this direction, since this, as noted above, is necessary for consistent definitions of T and μ. In a second set of studies, uniform systems are prepared at densities ρ_L and ρ_V, and allowed to attain coexistence under weak global exchange. As shown in Fig. 3, the coexisting densities in the pair of uniform systems are quite different from those observed in the single phase separated system. Studies of phase separation in another nonequilibrium lattice gas (with different sublattices in contact with reservoirs at distinct temperatures) yield similar results [14]. We are thus led to the conclusion that phase coexistence between NESS is not well defined: the coexisting densities depend not only on the drive and the reservoir temperature(s), but on the details of how the two phases make contact and exchange particles.

5. Conclusions

Summarizing, we have seen that steady-state thermodynamics appears viable for spatially uniform nonequilibrium steady states. It appears to fail, however, in the presence of nonuniformities, due either to a nonuniform drive, or to phase separation. This is an area in which simulation can help to answer fundamental questions regarding the scope

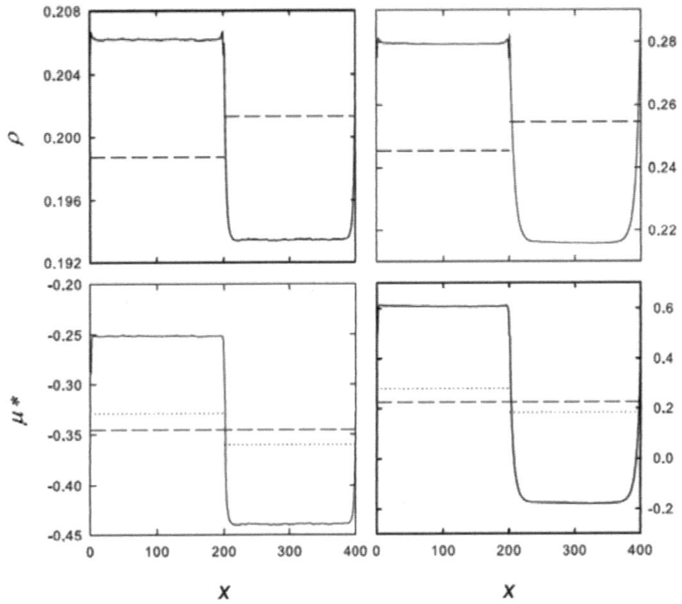

Figure 2. Density (upper) and chemical potential (lower) profiles in the half-driven NNE model. Drive $D = 0$ for $x \leq 200$; $D = 1$ for $x > 200$. Global densities $\overline{\rho} = 0.2$ (left) and 0.25 (right). In the upper panels the dashed lines show the coexisting densities predicted by equating chemical potentials. In the lower panels the dashed lines show the expected uniform value of the chemical potential, and the dotted lines show the chemical potentials for the isolated systems, each at density $\overline{\rho}$.

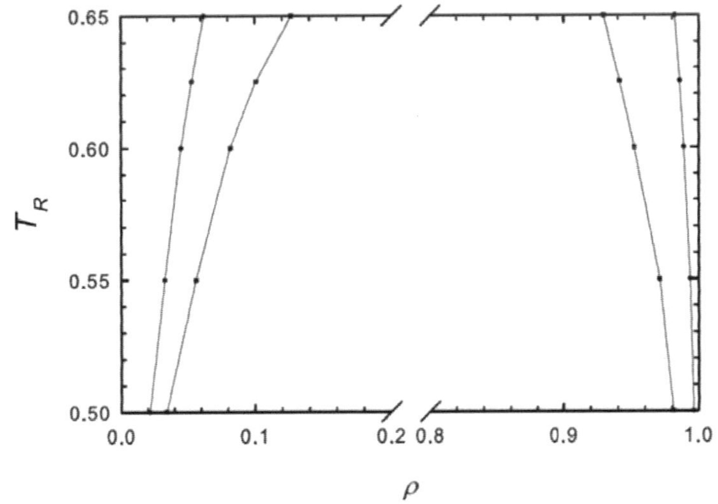

Figure 3. KLS model: coexisting densities ρ_L (right) and ρ_V (left) versus reservoir temperature T_R. The points nearer the vertical axes correspond to the single phase-separated system; those further from the axes correspond to uniform systems under weak global exchange. System size $L = 400$. Error bars smaller than symbols.

of thermodynamics. Further studies, of off-lattice systems, and of systems of particles with momentum, are needed to assess the validity of SST under more realistic conditions. Nevertheless, given the inconsistencies found in the simple models discussed here, it appears unlikely that a consistent thermodynamics of far-from-equilibrium systems is possible.

Acknowledgments
I thank Royce K. P. Zia for helpful comments. This work was supported by CNPq and CAPES, Brazil.

References

[1] Callen H B *Thermodynamics and an Introduction to Thermostatistics*, 2nd ed. (Wiley, New York, 1985)
[2] Onsager L 1931 Phys. Rev. **37** 405; **38** 2265
[3] de Groot S R and Mazur P *Non-Equilibrium Thermodynamics* (North-Holland, Amsterdam, 1962)
[4] Oono Y and Paniconi M 1998 *Prog. Th. Phys. Supp.* **130** 29
[5] Sasa S and Tasaki H 2006 *J. Stat. Phys.* **125** 125
[6] Hayashi K and Sasa S 2003 *Phys. Rev. E* **68** 035104
[7] Pradhan P, Ramsperger R, and Seifert U 2011 *Phys. Rev. E* **84** 041104
[8] Dickman R and Motai R 2014 *Phys. Rev. E* **89** 032134
[9] Dickman R 2014 *Phys. Rev. E* **90** 062123
[10] Widom B 1963 *J. Chem. Phys.* **39** 2808
[11] Guo W and Blöte H W J 2002 *Phys. Rev. E* **66** 046140
[12] Dickman R 2001 *Phys. Rev. E* **64** 016124
[13] Katz S, Lebowitz J L and Spohn H 1983 *Phys. Rev. B* **28** 1655
[14] Dickman R, arXiv: 1602.03507

An answer checking method for quantum annealers

Dilina Perera[1], M.A. Novotny[1,2,3]

[1] HPC[2] Distributed Analytics and Security Institute, Mississippi State University, Mississippi State, MS 39762-5167, USA

[2] Department of Physics and Astronomy, Mississippi State University, Mississippi State, MS 39762-5167, USA

[3] HPC[2] Center for Computational Sciences, Mississippi State University, Mississippi State, MS 39762-5167, USA

E-mail: dilina@dasi.msstate.edu

Abstract. We present a generic approach for checking the validity of the solutions returned by quantum annealing devices to aid in the analysis of whether the solution is the true ground state of the desired problem. The underlying principle is to embed a mirrored graph G' of the original graph G, and connect the two graphs via strong ferromagnetic/antiferromagnetic couplings that span across the mirror plane. This allows one to dismiss solutions that do not agree with the underlying mirror symmetry inherent to the true ground state of the composite graph. Using a 1000 qubit D-Wave 2X device, we demonstrate this method by applying it to spin glass problems defined on the device's native Chimera architecture.

1. Introduction

Combinatorial optimization [1] represents a generic class of problems that has direct applications in a broad array of fields ranging from manufacturing and transportation to machine learning. A vast majority of combinatorial optimization problems, including the boolean satisfiability (k-SAT) problem, the traveling salesman problem, knapsack problem etc. can be formulated as quadratic unconstrained binary optimization (QUBO) problems [2]. Each QUBO problem, in turn, can be mapped onto an Ising spin glass [3, 4] of the form

$$\mathcal{H}_{\text{prob}} = -\sum_{(i,j)\in E} J_{i,j} s_i s_j - \sum_{i\in V} h_i s_i \qquad s_i \in \{\pm 1\}, \tag{1}$$

where the graph $G = (V, E)$ with vertices V and edges E captures the structural information pertaining to the problem, with J_{ij} and h_i representing the exchange couplings and the local fields, respectively. Finding the ground state of the Hamiltonian (1) through exact enumeration is a non-deterministic polynomially (NP) hard problem [5], and hence in practice it is tackled via heuristic methods such as simulated annealing (SA) [6].

A relatively new optimization scheme called quantum annealing (QA) [7, 8, 9, 10] has recently gained widespread attention, particularly with the introduction of a line of commercial programmable devices by the D-Wave Systems Inc. [11] which utilize QA at the hardware level [12]. Unlike SA which circumvents local minima by thermal hopping over energy barriers, QA strives to avoid local minima with the aid of artificially introduced quantum fluctuations that allow the system to tunnel though energy barriers. These quantum fluctuations are introduced by mapping the classical Ising spins s_i in the problem Hamiltonian to Pauli z-matrices σ_i^z and incorporating a transverse magnetic field in the x direction, which leads to the time-dependent Hamiltonian

$$\mathcal{H}(t) = -A(t)\sum_i \sigma_i^x + B(t)\mathcal{H}_{\text{prob}}, \tag{2}$$

where the parameters $A(t)$ and $B(t)$ represent the annealing schedule. $A(0) \gg B(0)$, and hence at $t = 0$, the transverse field term dominates the Hamiltonian. At $t = 0$, the system is prepared in the ground state of the transverse field, which is an equal superposition state $(1/\sqrt{2})^n \prod_i(|\uparrow\rangle + |\downarrow\rangle)$ of all n qubits in the system. The goal of QA is to gradually decrease $A(t)$ while increasing $B(t)$ so that at the end of the annealing time $t = \tau$, $\mathcal{H}_{\text{prob}}$ dominates the Hamiltonian, i.e. $\mathcal{H}(\tau) \simeq \mathcal{H}_{\text{prob}}$. If the annealing is performed sufficiently slowly, the time evolution of the system would follow the instantaneous ground state of $\mathcal{H}(t)$ at all $t \in [0, \tau]$, which would ultimately result in the ground state of $\mathcal{H}_{\text{prob}}$ at $t = \tau$.

In August 2015, D-Wave Systems Inc. announced the breaking of the 1000 qubit barrier with the release of their latest quantum annealing device, the D-Wave 2X [13]. The device operates at a temperature of about 15 mK, with the annealing time in the range of 20 – 2000 μs. In D-Wave devices, the connectivity between the qubits is constrained to the so-called Chimera graph (See Fig. 1 for an illustration of a 4 × 4 Chimera graph.). A Chimera unit cell is a $K_{4,4}$ complete bipartite graph with each of the four qubits on the left/right connected to all the four qubits on the right/left. In addition, each qubit on the left is vertically connected to the corresponding qubit of the unit cells directly above and below, while each qubit on the right is coupled horizontally with the respective qubit of the unit cells directly to the left and to the right. The D-Wave 2X utilizes a 12 × 12 chimera graph, resulting in a total number of 1152 physical qubits, among which some of them are inaccessible to the users. This study used the device located at the NASA Ames Research Center which consists of 1097 accessible qubits.

Just as in any software-based heuristic method, a QA device may not necessarily find the global energy minimum for a problem during a given annealing run. In this paper, we present an "answer checking method" that allows one to perform one test to assess the validity of the solution as the true ground state of the problem. In section 2, we provide a detailed description of our technique, along with its limitations. In section 3, we demonstrate the application of this method to generally allowed spin glass problems on the D-Wave 2X device's native Chimera topology.

2. Answer checking via mirrored graphs

Our answer checking method exploits the symmetry conditions associated with the ground state solutions of graphs with mirror symmetry. The steps involved in the method can be listed as follows:

- Split the native architecture of the annealing device into two identical halves by defining a mirror axis.
- Embed the desired graph G onto one side of mirror axis.
- Embed the mirror image (G') of the original graph G onto the opposite side of the mirror axis.
- Connect a selected set of qubits of graph G with the corresponding mirror qubits on G' via strong ferromagnetic/antiferromagnetic couplings that span across the mirror axis (Henceforth denoted as "mirror couplings".).

Fig. 1 illustrates the application of this method to a hypothetical QA device with a 4 × 4 native Chimera architecture. Note that in order to make the two halves of the device identical, one should mirror any inaccessible qubits and couplings present in the device onto the respective sides of the mirror axis prior to embedding the desired graphs.

Let $q_i \in G$ and $q_i' \in G'$ respectively denote a chosen qubit and the corresponding mirror qubit that are connected via a mirror coupling M. Since the energy associated with this interaction takes the form $E = -Mq_iq_i'$, during the annealing process, a sufficiently large value of $|M|$ will impose the constraint $q_i = q_i'$ for the case of ferromagnetic coupling ($M > 0$), and $q_i = -q_i'$ for the case of antiferromagnetic coupling ($M < 0$). To understand how these constraints imposed by the mirror couplings can assist one to validate solutions, let us first consider the simplest scenario in which the ground state of the original graph G is non-degenerate. If the device succeeds at finding the true ground state of the composite graph $G_T = G + G'$, for $M > 0$, the corresponding classical spin configuration will possess reflection symmetry with respect to the mirror axis. For $M < 0$, the solution will satisfy spin-flip symmetry, i.e. if the spins on one side of the mirror are inverted, the resulting spin configuration will possess reflection symmetry about the mirror plane. If the ground state of the original graph G is degenerate, in principle, it is possible to have ground state solutions of G_T that are comprised of two different ground state configurations of G respectively occupying the two sides of the mirror plane, which also happens to satisfy the constraints imposed by the mirror couplings. In such cases, the solution will not satisfy the aforementioned symmetry conditions. However, the probability of such asymmetric ground state solutions significantly reduces as the ground state degeneracy of G decreases. Moreover, the spatial correlations induced by the mirror couplings increases the likelihood of occurrence of symmetric solutions during the annealing process.

The arguments presented above suggest that the symmetry of the spin configuration with respect to the mirror plane provides a means of validating the solutions returned by the device. However, it should be noted that the presence of reflection/spin-flip symmetry does not necessarily guarantee that the true ground state of G_T has been obtained, since the occasional occurrence of an excited state of G and its mirror counterpart respectively occupying the two sides of the mirror plane would also satisfy the same symmetry conditions. Therefore, the presence of symmetry should rather be regarded as a heuristic test that increases one's expectation that the true ground state has been realized.

3. Results

Here we demonstrate the application of the answer checking method to Ising spin glass problems defined on the device's native Chimera topology. The 12×12 Chimera graph of the D-Wave 2X was divided into two 12×6 subgraphs via a vertical mirror axis located between the 6th and 7th columns of unit cells. The horizontal couplings that connect these two columns of unit cells were designated as ferromagnetic mirror couplings, with the magnitudes set to the maximum allowed value of 1. As the problem graph G, we considered Chimera graphs with different sizes, for which we varied the number of columns of unit cells (N) from 1 to 6 while keeping the number of rows fixed to 12. First, the local fields h_i were set to zero, and the couplings were randomly chosen from three different instance classes: $J_{ij} \in \{\pm 1\}$ (U_1 class), $J_{ij} \in \{\pm 5/7, \pm 6/7, \pm 1\}$ ($U_{5,6,7}$ class), and $J_{ij} \in \{\pm 8/28, \pm 13/28, \pm 19/28, \pm 1\}$ (S_{28} class) [14, 15]. From each class, 1000 random instances were selected. For each of these instances, we performed 1000 annealing runs and examined the solution/solutions that correspond to the lowest energy. If at least one of the solutions that correspond to the lowest energy was found to be symmetric with respect to the mirror plane, it was considered as an indication that the true ground state might have been achieved. Based on the results of the 1000 random instances in each class, we calculated the probability P_{sym} that at least one of the solutions corresponding to the lowest energy is found to be symmetric. Fig. 2 (a) shows P_{sym} as a function of N for the three instance classes. Since the number of qubits in the problem is directly proportional to N, for all three instance classes, we observe a rapid decrease in P_{sym} with increasing N. Fig. 2 (b) demonstrates the influence

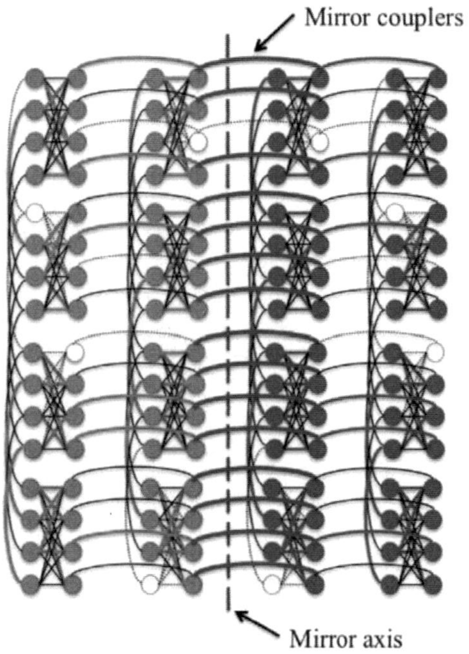

Figure 1. Application of the answer checking method to a hypothetical quantum annealing device with a 4×4 native Chimera architecture. The native architecture is divided into two identical halves via the vertical mirror axis. White-colored qubits and the couplings marked as dotted lines represent device's inaccessible qubits and couplings and their mirror counterparts. The original graph (red) is embedded onto the left half of the device while its mirror image (blue) is embedded onto the right half. The horizontal couplings that connect the 2nd and the 3rd columns of unit cells in the device are designated as mirror couplings (green).

of the random local fields h_i on P_{sym}. Here, we have compared the results obtained for the U_1 class with/without the presence of local fields randomly chosen to be either $+1$ or -1. Since non-zero local fields act as biases to the spins and render the problem easier to solve, we observe an increase in P_{sym} with the inclusion of the fields.

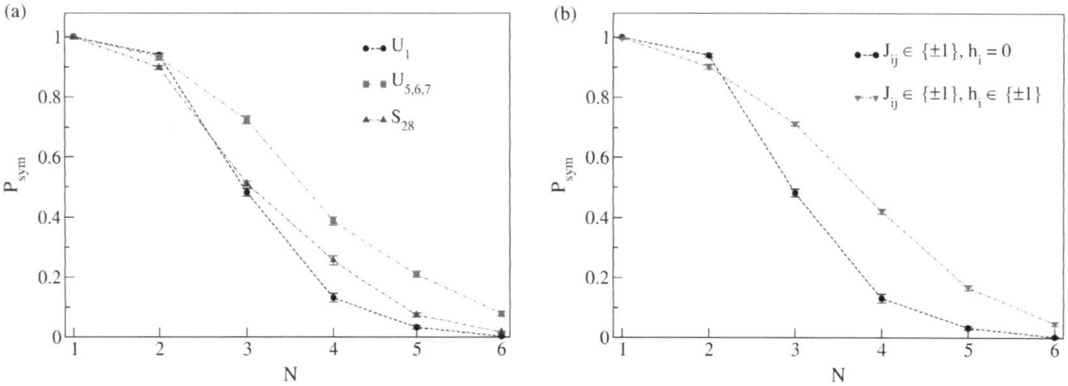

Figure 2. Probability (P_{sym}) that out of 1000 annealing runs, at least one of the solutions corresponding to the lowest energy is found to be symmetric with respect to the mirror plane. Number of columns of unit cells (N) in the problem graph was varied from 1 to 6 while keeping the number of rows fixed to 12. (a) P_{sym} vs. N for three different instance classes with zero local fields: $J_{ij} \in \{\pm1\}$ (U_1 class), $J_{ij} \in \{\pm5/7, \pm6/7, \pm1\}$ ($U_{5,6,7}$ class), and $J_{ij} \in \{\pm8/28, \pm13/28, \pm19/28, \pm1\}$ (S_{28} class). (b) P_{sym} vs. N for the U_1 class with/without random local fields $h_i \in \{\pm1\}$.

4. Summary

We have introduced an answer checking method for quantum annealers that enhances one's expectation that the true ground state of the problem has been obtained. The method relies on symmetry requirements associated with the ground state solutions of graphs with mirror symmetry. Our approach has some of the same flavor as a recently introduced quantum annealing error correction scheme [16, 17] in that both require duplicating the original problem. However, in contrast to the aforementioned error correction scheme in which each qubit of the original and the duplicated graphs are ferromagnetically coupled to a penalty qubit, in our approach, a selected set of qubits of the original graph are directly coupled to the corresponding mirror qubits. Our method is independent of the details of the device's native topology, and hence is readily applicable to any quantum annealing device with an arbitrary native graph structure.

Acknowledgments

This research was sponsored by Pacific Northwest National Laboratory (PNNL). Time on the D-Wave 2X was granted through Universities Space Research Association (USRA).

References

[1] Papadimitriou C H and Steiglitz K 1998 *Combinatorial Optimization: Algorithms and Complexity* (New York: Dover Publications)
[2] Lucas A 2014 *Front. Phys.* **2** 5
[3] Binder K and Young A P 1986 *Rev. Mod. Phys.* **58** 801
[4] Nishimori H 2001 *Statistical Physics of Spin Glasses and Information Processing: An Introduction* (New York: Oxford University Press, New York)
[5] Barahona F 1982 *J. Phys. A* **15** 3241
[6] Kirkpatrick S, Gelatt C D and Vecchi M P 1983 *Science* **220** 671
[7] Finnila A, Gomez M, Sebenik C, Stenson C and Doll J 1994 *Chem. Phys. Lett.* **219** 343
[8] Kadowaki T and Nishimori H 1998 *Phys. Rev. E* **58** 5355
[9] Santoro G E, Martoňák R, Tosatti E and Car R 2002 *Science* **295** 2427
[10] Santoro G E and Tosatti E 2006 *J. Phys. A* **39** R393
[11] `http://www.dwavesys.com`
[12] Johnson M W, Amin M H S, Gildert S, Lanting T, Hamze F, Dickson N, Harris R, Berkley A J, Johansson J, Bunyk P, Chapple E M, Enderud C, Hilton J P, Karimi K, Ladizinsky E, Ladizinsky N, Oh T, Perminov I, Rich C, Thom M C, Tolkacheva E, Truncik C J S, Uchaikin S, Wang J, Wilson B and Rose G 2011 *Nature* **473** 194
[13] Announcing the D-Wave 2X quantum computer URL `http://www.dwavesys.com/blog/2015/08/announcing-d-wave-2x-quantum-computer`
[14] Katzgraber H G, Hamze F, Zhu Z, Ochoa A J and Munoz-Bauza H 2015 *Phys. Rev. X* **5** 031026
[15] Zhu Z, Ochoa A J, Schnabel S, Hamze F and Katzgraber H G 2016 *Phys. Rev. A* **93** 012317
[16] Pudenz K L, Albash T and Lidar D A 2014 *Nat. Commun.* **5** 3243
[17] Pudenz K L, Albash T and Lidar D A 2015 *Phys. Rev. A* **91** 042302

Journal of Physics: Conference Series **750** (2016) 012006 doi:10.1088/1742-6596/750/1/012006

Stable knots in the phase diagram of semiflexible polymers: A topological order parameter?

Wolfhard Janke and Martin Marenz

Institut für Theoretische Physik, Universität Leipzig, Postfach 100 920, 04009 Leipzig, Germany

E-mail: wolfhard.janke@itp.uni-leipzig.de, martin.marenz@itp.uni-leipzig.de

Abstract. We report a recent computer simulation study of the phase diagram of a single semiflexible polymer chain using a generic coarse-grained bead-stick model. The results are obtained with a combination of multicanonical and replica-exchange simulations. Interestingly, we find in the temperature-stiffness plane well-defined phases that can be characterized by stable knots of various types. We discuss how the knots can be observed and to what extent they can be interpreted as "topological order parameter". Another surprising observation concerns the peculiar transitions into these stable knot phases which will be also explained.

1. Introduction

The statistical physics of polymer systems is a very rich field with countless applications ranging from nanotechnology over molecular biology to all kinds of plastics. By stochastically sampling the state space, Monte Carlo (MC) computer simulations provide thermodynamic and structural information of the system. The latter is particularly important for polymers whose typically linear structure gives rise to many different structural motifs [1]. A good example is the phase diagram of a single semiflexible polymer as a function of temperature and bending stiffness [2, 3]. Using a coarse-grained bead-stick continuum model we recently scanned the full range from flexible to stiff polymers and determined the transition lines between phases characterized by well-known structural motifs such as rod-like, collapsed, frozen, bent, and hairpin structures. Among them we also discovered peculiar phases governed by stable knots of various types [3]. Here we give an overview of this study, explain how knots can be detected and why they may be interpreted as "topological order parameter". Finally we also discuss the rather intriguing properties of the transitions into these stable knot phases.

2. Model and simulation methods

In Ref. [3] we considered a minimalistic coarse-grained bead-stick model of a linear semiflexible polymer consisting of N monomers connected by fixed bonds of length r_b. Two terms contribute to the energy [4]:

$$E = E_{\rm LJ} + \kappa E_{\rm Bend} = 4\epsilon \sum_{i=1}^{N-2} \sum_{j=i+2}^{N} \left[\left(\frac{\sigma}{r_{ij}}\right)^{12} - \left(\frac{\sigma}{r_{ij}}\right)^{6} \right] + \kappa \sum_{i=1}^{N-2} (1 - \cos\vartheta_i) \; . \qquad (1)$$

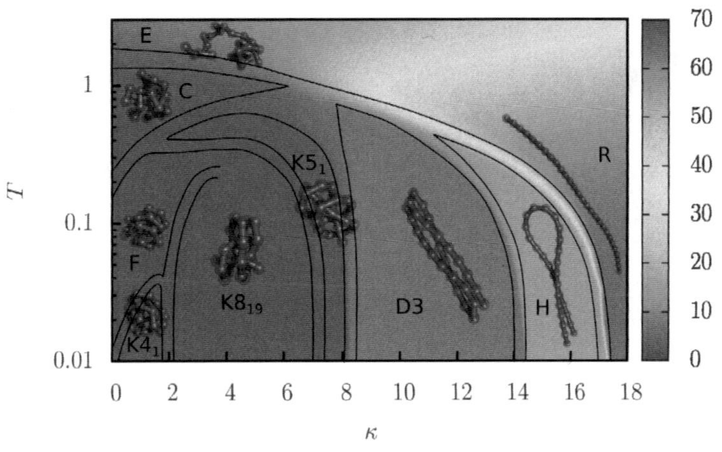

Figure 1. Phase diagram of a semiflexible 28mer in the temperature (T) – bending stiffness (κ) plane (E: elongated, R: rod-like, C: collapsed, F: frozen, K: knotted, DN: ($N-1$) times bent, H: hairpin). The background color encodes $\langle R^2_{\rm gyr}\rangle$. Note that the temperature is given on a logarithmic scale (taken from Ref. [3]).

The first term is the standard 12–6 Lennard-Jones (LJ) potential, where r_{ij} measures the distance between the monomers i and j. The parameters ϵ and σ set the energy and length scales, respectively, and hence are taken to be $\epsilon = 1$ and $\sigma = 1$ in the following. Note that the minimum of the individual contributions to the LJ potential occurs at $r_{\rm LJ} = 2^{1/6} \approx 1.12 > r_b = 1$. The second term models the bending energy of the polymer similar to a worm-like chain, where $0 \leq \vartheta_i \leq \pi$ is the angle between adjacent bonds and the parameter κ controls the stiffness strength. For $\kappa = 0$ the polymer is flexible and for large κ it approaches the rod-like limit.

In order to map out the phase diagram in the $T - \kappa$ plane we performed two types of MC studies in generalized ensembles. In the first, we employ a combination of multicanonical simulations with replica exchange in κ direction, and the second is based on a two-dimensional variant of replica exchange in T and κ. In multicanonical ("muca") simulations [5, 6, 7] one replaces the usual Boltzmann weight e^{-E/k_BT} in the canonical partition function by an *a priori* unknown weight function $W(E)$ for polymer configurations x with energy $E(x)$. This results in $Z_{\rm muca} = \sum_x W(E(x)) = \sum_E \Omega(E)W(E)$, where $\Omega(E)$ is the density of states. One usually aims at adjusting $W(E)$ such that the multicanonical energy histogram $H(E) \propto P_{\rm muca}(E) = \Omega(E)W(E)$ becomes approximately flat. This can be achieved by starting with $W^{(0)}(E) \equiv 1$ and iterating $W^{(n+1)}(E) = W^{(n)}(E)/H^{(n)}(E)$, where $H^{(n)}(E)$ is the simulated histogram using the weight $W^{(n)}(E)$. The basic idea of replica exchange (or parallel tempering) [8] is to simulate m replicas at different temperatures $k_BT_\mu = 1/\beta_\mu$, $\mu = 1, \ldots, m$, and to propose every now and then an exchange of the configurations x of two replicas μ and ν. To ensure detailed balance, these exchanges are accepted with probability $p(x_\mu \leftrightarrow x_\nu) = \min\{1, \exp(\Delta\beta\Delta E)\}$, where $\Delta\beta = \beta_\mu - \beta_\nu$ and $\Delta E = E(x_\mu) - E(x_\nu)$ are the differences in the inverse temperatures and energies of the two replicas. For a more comprehensive discussion of state-of-the-art simulation and analysis methods in polymer physics, see the recent review [1].

3. Results

The resulting pseudophase diagram for a semiflexible 28mer as obtained in Ref. [3] is shown in Fig. 1. The transition lines were determined by measurements of the two subenergies $\langle E_{\rm LJ}\rangle$

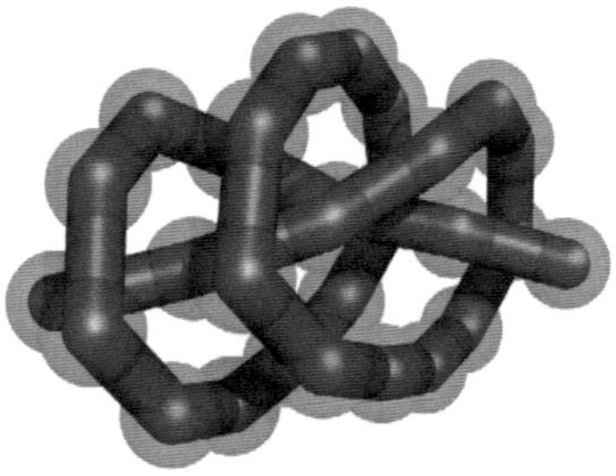

5_1

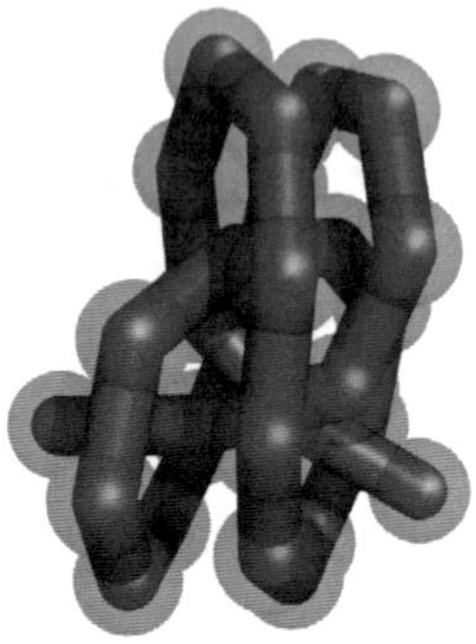

8_{19}

Figure 2. Typical knots of types 5_1 (at $\kappa = 7.50$, $T = 0.045$) and 8_{19} (at $\kappa = 6.10$, $T = 0.035$) obtained for a semiflexible 28mer.

and $\langle E_{\mathrm{Bend}} \rangle$, the squared end-to-end distance $\langle R_{ee}^2 \rangle$, the squared radius of gyration $\langle R_{\mathrm{gyr}}^2 \rangle$, and the eigenvalues of the gyration tensor $\langle \lambda_1 \rangle$, $\langle \lambda_2 \rangle$, and $\langle \lambda_3 \rangle$. Peaks of the temperature derivative of these observables for a given bending stiffness κ then mark the transition lines in Fig. 1. Due to the finite length of the polymer, different observables give slightly different transition temperatures, which is reflected by the width between the black lines. The background color encodes the average extension of the polymer in terms of $\langle R_{\mathrm{gyr}}^2 \rangle$. We confirmed the thus obtained phase diagram by a microcanonical analysis [9, 10, 11, 12] where one starts from the density of states $\Omega(E)$, calculates the microcanonical entropy $S(E) = k_B \ln \Omega(E)$ and temperature $\beta_{\mathrm{micro}} = dS(E)/dE$, and finally analyzes the peaks in the derivative of $d\beta_{\mathrm{micro}}/dE$ [13].

For high temperatures, the typical conformations are extended (E) or rod-like (R). With decreasing temperature, depending on the bending stiffness, collapsed (C), frozen (F), bent (D3) and hairpin (H) conformations develop. Interestingly, we also observe phases governed by stable knots (K). Two examples are depicted in Fig. 2. Closer inspection reveals that these knots can be identified as $C_n = 4_1$, 8_{19}, and 5_1 knots, where according to the usual classification scheme C counts the minimal number of crossings of any projection of a knot onto a two-dimensional plane and the subscript n distinguishes topologically different knots with the same C. Note that the 5_1 and 8_{19} knots are so-called torus knots, which are known to be formed preferentially in viral DNA [14]. For the identification of the knot type, we employed a method described in Ref. [15] where a specific product $\Delta_p(t) \equiv |\Delta(t) \times \Delta(1/t)|$ of the Alexander polynomial $\Delta(t)$ is evaluated at $t = -1.1$. For definitions and a detailed description of mathematical knot theory, see the book by Kauffman [16].

Mathematically, the identification of knots in *open* polymers is not well defined, so one first has to apply a (virtual) closure prescription. The easiest one would be to draw a virtual bond connecting the two termini of the polymer, but this direct closure results in spurious knots when the polymer is rather compact. In this work we employed the closure CI sketched in Fig. 3(a), which is inspired by tying a real knot. First one connects the two termini of the polymer by a straight line which is then extended in both directions. The two new virtual endpoints A' and B' are located far away from all monomers. A third virtual point C is created on the perpendicular bisector of the connecting line, again far away from all monomers. Finally the polymer can be closed via straight lines connecting C with A' and B', respectively.

For testing purpose we also implemented the second closure CII sketched in Fig. 3(b). Here one increases the bond length of the first and last bond to two new virtual termini A' and B', which are then connected by a straight line to close the polymer. The closures CI and CII give

Journal of Physics: Conference Series **750** (2016) 012006 doi:10.1088/1742-6596/750/1/012006

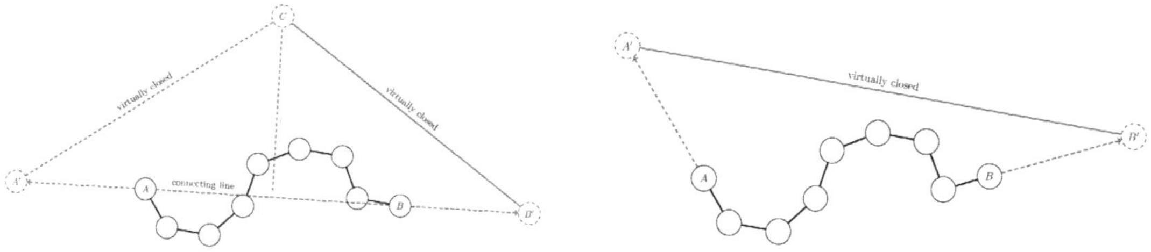

(a) (b)

Figure 3. (a) The closure CI moves the termini A and B of the polymer along the line connecting them far outside the polymer to points A' and B', which are then virtually closed by connecting them to a point C on the perpendicular bisector. (b) The closure CII first shifts the two termini A and B to points far outside the polymer and then closes the new virtual termini A' and B'.

qualitatively similar results and are both suitable for identifying the knotted regions. However, with CI all measured conformations are assigned an identical knot type, so $D \equiv \Delta_p(-1.1)$ can be considered as a topological order parameter.

In the study of Ref. [2] using a similar bead-spring model *no* knot phases have been reported. From our own test simulations we suspect that the reason for this difference lies in the choice of the ratio r_{LJ}/r_b, which is 1 in [2] but ≈ 1.12 in our case. If $r_{\mathrm{LJ}}/r_b \approx 1$, then bent conformations are energetically favored over knots. To verify this conjecture, however, more work is necessary.

The transitions from the frozen or bent phases into the knot phases are quite intriguing. As transitions between two structured states, one would expect them to behave first-order-like, similar to other solid-solid transitions. A glance at the inset of Fig. 4 for the D3–K5_1 transition suggests, however, that this expectation may not be true since the energy distribution $p(E)$ exhibits only a single peak. There is no indication for the typical double-peak structure at a first-order-like phase transition and hence no signal of latent heat [17, 18]. The true nature of the transition is only revealed when one considers the *two-dimensional* energy distribution $p(E_{\mathrm{LJ}}, E_{\mathrm{Bend}})$, for which indeed two clearly separated peaks are visible in Fig. 4 [3]. The peak in front corresponds to the (unknotted) bent phase D3 and the other in the back to the knot phase K5_1. The total energy $E = E_{\mathrm{LJ}} + \kappa E_{\mathrm{Bend}}$ is the projection along the diagonal of this two-dimensional histogram along which the two peaks fall on top of each other, what explains why only a single peak shows up in $p(E)$ and no latent heat is observable.

4. Summary

Using a combination of multicanonical and replica-exchange simulations, we mapped out the phase diagram for a generic bead-stick model of a semiflexible polymer. Besides the structural motifs already observed in previous work, we found transitions into novel phases governed by thermodynamically stable knots of various types, which may be considered as a topological order parameter. Their properties are considerably different from those of knots observed in the swollen and globular phases of flexible polymers, which form by chance. The second intriguing observation is that the transitions into these knotted conformations from other structured states happen with almost no latent heat, although we observed a clear phase coexistence.

Acknowledgments

This work was supported by the European Union and the Free State of Saxony, the Leipzig Graduate School of Natural Sciences "BuildMoNa", the Deutsche Forschungsgemeinschaft through Sonderforschungsbereich/Transregio SFB/TRR 102 (Project B04), and the Deutsch-

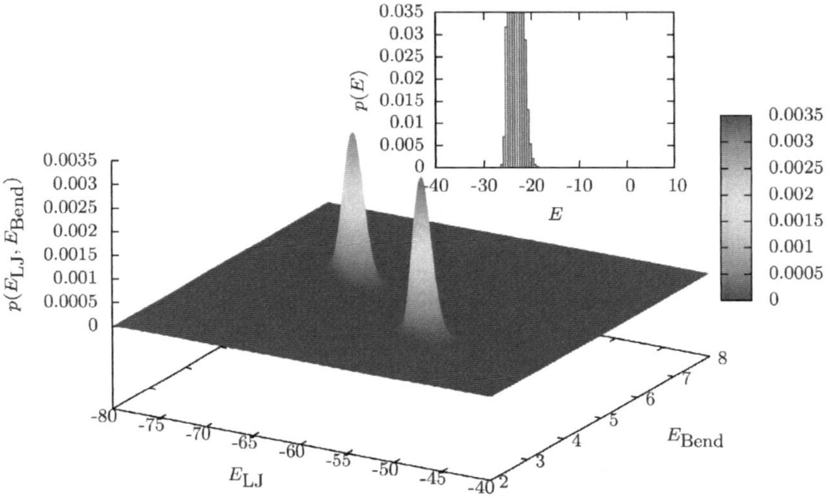

Figure 4. Two-dimensional energy histogram $p(E_{\mathrm{LJ}}, E_{\mathrm{Bend}})$ of a 28mer at the D3–K5_1 transition for $\kappa = 8.0$ at $T = 0.18$, signaling clear phase coexistence. The inset shows the one-dimensional energy histogram $p(E)$ of the total energy $E = E_{\mathrm{LJ}} + \kappa E_{\mathrm{Bend}}$, which corresponds to a projection along the diagonal of the two-dimensional histogram. In this projection, the two peaks fall on top of each other, so only a single peak is visible in $p(E)$.

Französische Hochschule (DFH-UFA) through the International (Leipzig-Lorraine-Lviv-Coventry) Doctoral College for the Statistical Physics of Complex Systems "$\mathbb{L}^4$" under Grant No. CDFA-02-07. Special thanks goes to the John von Neumann Institute for Computing (NIC) for granting computing time provided on the supercomputer JUROPA at Jülich Supercomputing Centre (JSC) under Grant No. HLZ21.

References

[1] Janke W and Paul W 2016 *Soft Matter* **12** 642-657
[2] Seaton D T, Schnabel S, Landau D P and Bachmann M 2013 *Phys. Rev. Lett.* **110** 028103-1-5
[3] Marenz M and Janke W 2016 *Phys. Rev. Lett.* **116** 128301-1-6
[4] Bachmann M, Arkın H and Janke W 2005 *Phys. Rev.* E **71** 031906-1-11
[5] Berg B A and Neuhaus T 1991 *Phys. Lett.* B **267** 249-253
[6] Berg B A and Neuhaus T 1992 *Phys. Rev. Lett.* **68** 9-12
[7] Janke W 1992 *Int. J. Mod. Phys.* C **3** 1137-1146
[8] Hukushima K and Nemoto K 1996 *J. Phys. Soc. Jpn.* **65** 1604-1608
[9] Gross D H E 2001 *Microcanonical Thermodynamics* (Singapore: World Scientific)
[10] Janke W 1998 *Nucl. Phys.* B (Proc. Suppl.) **63A-C** 631-633
[11] Junghans C, Bachmann M and Janke W 2006 *Phys. Rev. Lett.* **97** 218103-1-4
[12] Junghans C, Bachmann M and Janke W 2008 *J. Chem. Phys.* **128** 085103-1-9
[13] Schnabel S, Seaton D T, Landau D P and Bachmann M 2011 *Phys. Rev.* E **84** 011127-1-4
[14] Reith D, Cifra P, Stasiak A and Virnau P 2012 *Nucleic Acids Res.* **40** 5129-5137
[15] Virnau P 2010 *Physics Procedia* **6** 117-125
[16] Kauffman L H 1991 *Knots and Physics* 2nd ed. (Singapore: World Scientific)
[17] Janke W 1993 *Phys. Rev.* B **47** 14757-14770
[18] Janke W 2003 in: *Computer Simulations of Surfaces and Interfaces*, NATO Science Series, II. Mathematics, Physics and Chemistry – Vol. **114**, edited by B. Dünweg, D. P. Landau and A. I. Milchev (Dordrecht: Kluwer), pp. 111-135

Determination of Shape-Reactivity Relationships in Al-Nanoclusters

Sara E. Mason, Katie W. Corum, and Jennifer L. Bjorklund

Department of Chemistry, University of Iowa, Iowa City, Iowa 52242, USA

Corresponding author email address: sara-mason@uiowa.edu

Abstract. First-principles modeling of reactive materials offers a means of developing molecular-level understanding of adsorption processes. Geochemical surface science is in need of conceptual understanding of what substrate properties govern reactivity. Here, we review and expand upon some of our density functional theory modeling studies of ion adsorption on giant polyaluminum cation species. Where appropriate, we relate our findings to experimental information, and aim to generalize our results to enable future predictions.

1. Introduction

The partitioning of contaminant species between aqueous and solid phases influences bioavailability, transport, and fate in the environment. Fundamentally, ionic adsorption by solid phases is related to processes of importance to heterogeneous catalysis, electrochemistry, and corrosion. A reductionist approach to studying heterogeneous interface systems is to use well-defined simplified models, and then systematically increase complexity. In geochemical surface science, single-crystal surfaces have emerged as a popular model system, giving rise to more than a decade of experimental and theoretical work. [1-7] However, it is not obvious how to advance these models to take into account defects that better represent natural surfaces while still employing the experimental and computational methods that rely on surface order. An alternative approach to geochemical surface science is to employ well-defined molecular models, such as polyaluminum cations based on the Baker-Figgis-Keggin structural motif. Such aluminum hydroxide nanoclusters are attractive geochemical models because they can be found in nature, are readily synthesized in the lab, and, between their size and crystal structure information, are also tractable at the atomistic modeling level. [8-10] The cation Al_{13} ($Al_{13}O_4(OH)_{24}(H_2O)_{12}^{7+}$) has been identified in the δ- and ε-isomers, with the ε-isomer being the most common form found in nature. Upon heating, the δ-Al_{13} cluster has been shown to form even larger complexes including Al_{30} ($Al_{30}O_8(OH)_{56}(H_2O)_{26}^{18+}$) and Al_{32} ($Al_{32}O_8(OH)_{60}(H_2O)_{30}^{20+}$), where two δ-$Al_{13}$ are connected with Al^{3+} linkers. [11,12]

Arsenate and phosphate polyoxoanions share many properties and are important contaminants to target in water purification because of their impact on human health. In the environment, P(V) has become of greater concern because of phosphorous enrichment in fertilizers. Because it is toxic and a known carcinogen, As water contamination is a major health concern worldwide. Speciation of As and P in the environment results in a variety of derivatives that are also of concern. [13-15] Recently, the arsenic uptake capacity of Al-nanoclusters was compared to other Al-based sorbents, including amorphous polyaluminum chloride and mineral Al phases. [16] It was shown that the Al-nanoclusters are the most efficient for arsenic removal from water. Polyaluminum cations are used as coagulants and adsorbents in water remediation processes. A number of studies have shown that 1 nm Al_{13} clusters perform better than alum ($KAl(SO_4)_2 \cdot nH_2O$) as a clarifying agent, and 2 nm Al_{13} performs even better over a wider range of pH and dosage. [16-19]

The growing empirical evidence of desirable adsorption properties possessed by Al-nanoclusters has motivated an ongoing collaborative "make and model" approach towards understanding the reactivity of this important geochemical model/water treatment sorbent. Here, we review and expand upon our recent studies, and discuss future research directions. An unexpected outcome of this body of work is that much of the conventional wisdom about metal hydroxide surface reactivity is at odds with the results. To briefly review, geochemical surface science has relied on the concept of bond-valence to rationalize structure-reactivity relationships. The bond-valence model of Brown, parametrized using bulk structures, suggests that oxygen functional group coordination and protonation states are key predictors of metal (hydr)oxide surfaces. [20] There are also considerations of adsorption based on the type of interaction between the ion and the surface. In inner-sphere (IS) adsorption, the adsorbate forms direct bonds to the surface, while in outer-sphere (OS) adsorption, the interaction is predominated by electrostatic interactions. Based on the different bonding in IS and OS adsorption, it is generally assumed that OS adsorption is relatively weak. [21]

2. Computational methods

Molecular structures for Al_{13} and Al_{30} were based on crystal structures. [11,22] Counter ions and molecular water not bound to the particles were removed, and the protonation state of under-coordinated oxygen atoms was deduced based on bond-valence. The molecular geometries were optimized using density functional theory (DFT), embedded in the conductor-like screening model (COSMO) dielectric continuum model, as implemented in the $DMol^3$ code developed by Delley. [23,24] The DFT calculations are aperiodic, and include spin-polarization when adsorbates with an odd number of electrons are considered. The generalized gradient approximation (GGA) of Perdew, Burke, and Ernzerhof (PBE) is used, [25] and the single-electron wavefunctions are expanded in a double-numeric-plus polarization atom-centered basis set. Because there was no significant difference in total energies, reaction energies, or the magnitude of the forces, the real-space basis set cut off of 3.5 Å was used. Further details of the methods, including benchmarking of geometries and relative energies using different computational parameters, can be found in our published studies. [10,26-28]

3.1 Trends in ion adsorption on Al_{30}

In a purely theoretical study by Corum and Mason, [26] trends in cation and anion adsorption on Al_{30} were explored. Based on studies of Al mineral surfaces, the cations Cu^{2+}, Pb^{2+}, and Zn^{2+} were assumed to adsorb to the particle surface in an IS bidentate fashion, while anions SO_4^{2-} and Cl^- were assumed to adsorb in an OS fashion. Figure 1 summarizes the η-H_2O functional group distinction of Rustad, [29] which is based on the distance of bound water from tetrahedral Al atoms in each end of the symmetric molecule. For convenience, we also refer to the particle region between the two δ-Al_{13} building-blocks of Al_{30} as the beltway; the flat ends of the δ-Al_{13} sections are referred to as the caps.

Journal of Physics: Conference Series **750** (2016) 012007 doi:10.1088/1742-6596/750/1/012007

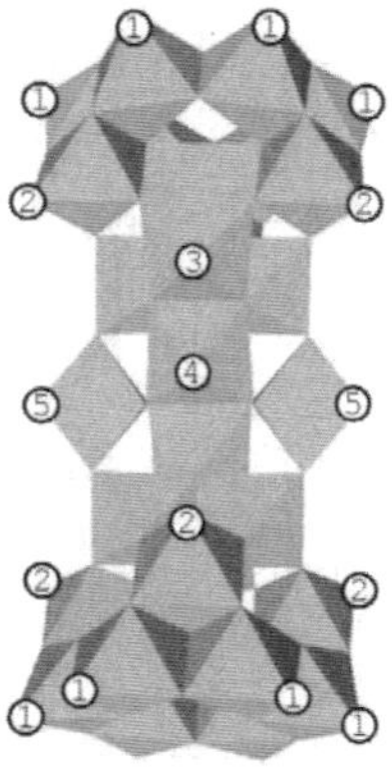

Figure 1. Blue polyhedral representation of ($Al_{30}O_8(OH)_{56}(H_2O)_{26}{}^{18+}$), showing the locations of bound η-H_2O groups. Taken from Figure 1a in Ref. [10]. Copyright American Chemical Society 2013.

Using DFT total energies and stoichiometric weights based on model adsorption reactions, we calculated theoretical adsorption energies, E_{rxn}. The results obtained show that Cu^{2+} and Pb^{2+} prefer adsorption through η-H_2O groups at the caps of Al_{30} by 1.04 and 1.13 eV, respectively. On the other hand, SO_4^{2-} and Cl^- prefer OS adsorption in the beltway region, by 1.22 and 1.10 eV, respectively. The fact that the IS adsorption varies by >1 eV between oxygen functional groups of identical coordination/protonation violates the bond-valence interpretation of surface reactivity, and the fact that anion adsorption exhibits an opposite site preference to cation adsorption suggests variation in reactivity as a function of location on the particle surface. We found that the electrostatic potential (V_{el}) shown in Figure 2, offers an explanation for this behavior: V_{el} is relatively negative on the molecular caps. On the other hand, in the beltway, the molecular shape gives rise to collective interactions of exposed water groups, leading to regions of relatively positive values of V_{el}. As the variation in V_{el} has less to do with function group coordination and more to do with topography, we term the governance of V_{el} on reactivity in Al_{30} as a "shape-reactivity relationship."

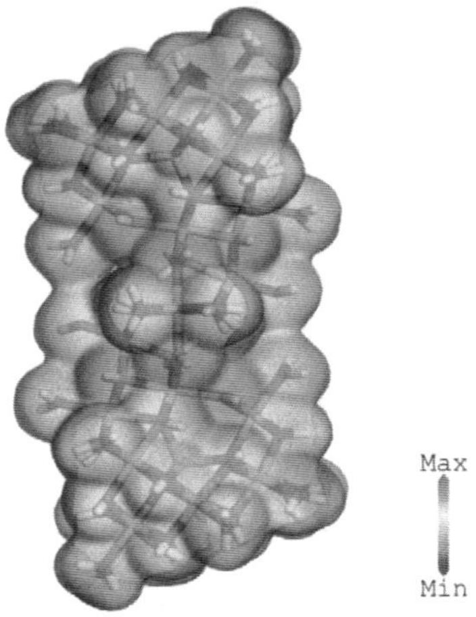

Figure 2. Electrostatic potential, V_{el}, of Al_{30} mapped onto a charge density isosurface. Taken from Figure 7 in Ref. [10]. Copyright American Chemical Society 2013.

Journal of Physics: Conference Series **750** (2016) 012007 doi:10.1088/1742-6596/750/1/012007

3.2 Co-adsorption of Cu^{2+} and SO_4^{2-} on Al_{30}

The experimental determination of the structure of Cu^{2+} bound to Al_{30} (in two symmetry equivalent positions) provided an opportunity to use our modeling approach to rationalize the observed adsorbed geometry of the cation. The experimental crystal structure, Cu_2Al_{30}-S, is shown in Figure 3, and exhibits Cu^{2+} in the beltway region, along with the counter ion used for crystallization, 2,6 napthalene disulfonate (2,6-NDS). Immediately it is apparent that Cu^{2+} adsorption in the presence of counterion conflicts with our theoretical result of Cu^{2+} preferring adsorption at the cap sites. To explore this theoretically, we compare the separate adsorption of IS Cu^{2+} and OS SO_4^{2-} as discussed above, and compare that to modeled co-adsorption. We consider sulfate adsorption in place of the more complicated counterion used in the experiment to simplify the calculations, and note that sulfate is also a suitable counterion for Al_{30} crystallization.

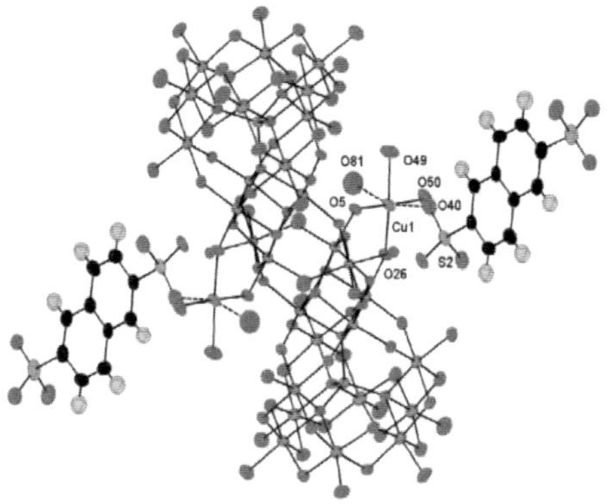

Figure 3. Optimized geometry of Cu_2Al_{30}-S, which shows the two Cu^{2+} cations near the beltway region with the sulfate group of the 2,6-NDS crystallizing agent. The color scheme for the Cu, Al, S, O, C, and H atoms are green, blue, orange, red, black, and gray, respectively. This atom labelling scheme is used throughout. Taken from Figure 3 in Ref. [10]. Copyright American Chemical Society 2013.

As reported in our collaborative study, [10] while copper and sulfate separately prefer the caps and beltway sites by 1.04 and 0.95 eV, respectively, the DFT energetics show that co-adsorption is preferred in the beltway by 1.10 eV, congruent with the Cu_2Al_{30}-S crystal structure. The opposing trends of isolated ions has already been explained here in terms of V_{el}. The co-adsorption result can be further understood by comparing the strength of the OS SO_4^{2-} adsorption compared to the IS Cu^{2+} adsorption. As noted in the Introduction, OS adsorption is generally thought to be weaker than IS adsorption, as the former is dominated by electrostatics while the latter involves direct bonding. However, the DFT values of E_{rxn} show that the most preferred value for IS Cu^{2+} adsorption is -1.26 eV, while that of OS SO_4^{2-} adsorption is -2.34 eV. This unusually strong OS SO_4^{2-} adsorption is explored by visualizing the adsorption-induced charge density $\Delta\rho_{SO4}$ taken as $\rho_{Al30/SO4}$ - ρ_{Al30} - ρ_{SO4}, shown in Figure 4. Qualitatively, it is apparent that there is a larger isosurface volume (at the same isosurface value) when sulfate adsorbs in the beltway (left) than at the caps (right). The plots of $\Delta\rho_{SO4}$ further support the concept of "shape-reactivity" relationships in Al_{30}: At the caps, the SO_4^{2-} -- Al_{30} interactions are sterically limited to a relatively small region of the particle surface, while the contours give rise to more extensive SO_4^{2-} -- Al_{30} interactions in the beltway.

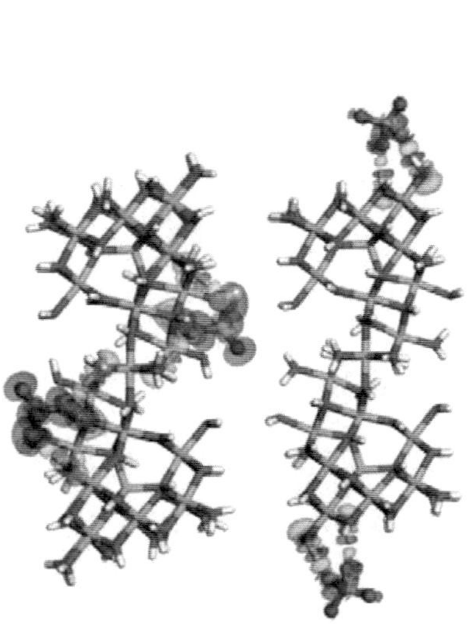

Figure 4. Induced charge density $\Delta\rho_{SO4}$ for the beltway and cap configuration, respectively. Charge loss is shown in black, and charge gain is shown in yellow. Adapted from Figure 6 Ref. [10], Copyright American Chemical Society, 2013.

Co-adsorption can be schematically summarized as shown in Figure 5: The topographical variations in V_{el} (left) give rise to opposing trends in isolated cation/anion adsorption site preference, with Cu^{2+} preferring the caps and SO_4^{2-} preferring the beltway (middle). However, the unusually strong OS SO_4^{2-} drives the site preference for co-adsorption in the beltway (right).

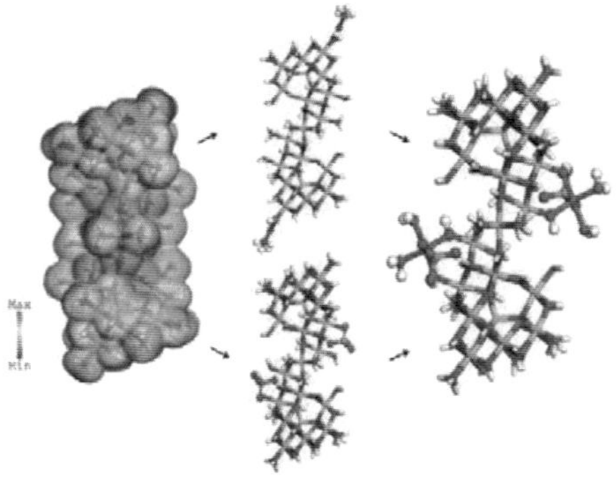

Figure 5. Schematic summary of coadsorption behavior of SO_4^{2-} and Cu^{2+} adsorption on Al_{30}, as described in the text.

3.3 As and P Adsorption on Al_{30}

More recently, we used the experimental crystal structure of $[(TBP)_2Al_2(\mu_4\text{-}O_8)(Al_{28}(\mu_2\text{-}OH)_{56}(H_2O)_{22})]^{14+}$ shown in Figure 6 (where TBP = *t*-butylphosphonate $(CH_3)_3CPO_3)$), as a basis for the DFT calculations to assess whether As(V) adsorption behavior follows that of P(V). We went on to explore if the beltway preference of the polyoxoanion forms of As(V)/P(V) could be changed through R-group substitution exploiting varying inductive effects. The R groups considered were *tert*-butyl (TB), methyl (M), benzene

(Ph), hydroxyl (OH), and methoxy (OCH_3). The Eigen-Wilkens (EW) mechanism [30,31] is shown in Scheme 1, where the net reaction is used to determine E_{rxn}.

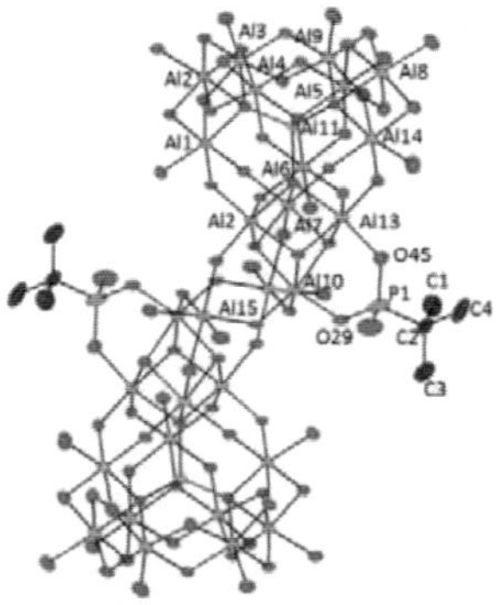

Figure 6. Experimental crystal structure of $[(TBP)_2Al_2(\mu_4\text{-}O_8)(Al_{28}(\mu_2\text{-}OH)_{56}(H_2O)_{22})]^{14+}$, where the Al^{3+} cations are represented as blue ellipsoids; O, P, and C atoms are red, orange, and black ellipsoids, respectively. Taken from Figure 2 in Ref. [27]. Copyright American Chemical Society 2015.

Scheme 1. Eigen-Wilkens mechanism that proceeds by the initial formation of an outer-sphere ion pair followed by ligand exchange to form the final product

Step 1: Ion Pair Formation

$$Al_{30}O_8(OH)_{56}(H_2O)_{22} - (OH)_2(H_2O)_2{}^{16+} + 2\,RXO_2OH^{1-}$$
$$\longrightarrow [Al_{30}O_8(OH)_{56}(H_2O)_{22} - (OH)_2(H_2O)_2{}^{16+}, 2\,RXO_2OH^{1-}]^{14+}$$

outer-sphere ion pair

Step 2: Ligand Exchange

$$[Al_{30}O_8(OH)_{56}(H_2O)_{22} - (OH)_2(H_2O)_2{}^{16+}, 2\,RXO_2OH^{1-}]^{14+}$$
$$\longrightarrow Al_{30}O_8(OH)_{56}(H_2O)_{22} - (RXO_2O)_2{}^{14+} + 4\,H_2O$$

Net Reaction

$$Al_{30}O_8(OH)_{56}(H_2O)_{22} - (OH)_2(H_2O)_2{}^{16+} + 2\,RXO_2OH^{1-}$$
$$\longrightarrow Al_{30}O_8(OH)_{56}(H_2O)_{22} - (RXO_2O)_2{}^{14+} + 4\,H_2O$$

The values of E_{rxn} show that the beltway configuration, congruent with the experimental crystal structure, is preferred by 0.42 eV for TBP and 0.51 eV for the arsenate analog. Ranges of 0.25 eV and 0.49 eV in E_{rxn} were obtained as a function of R-group for P(V) and As(V) species respectively, with the most electron donating group (TB) showing the most favorable values and the most electron withdrawing group (OCH_3) showing the least favorable values. However, while the magnitude of E_{rxn} was tunable in a fashion that followed predictions, it was always the case that the least favorable beltway adsorption of As or P was more favored than the most favorable cap adsorption of As or P. Thus we conclude that the beltway site preference can persist even as a function of large variation in arsenic or phosphorous speciation.

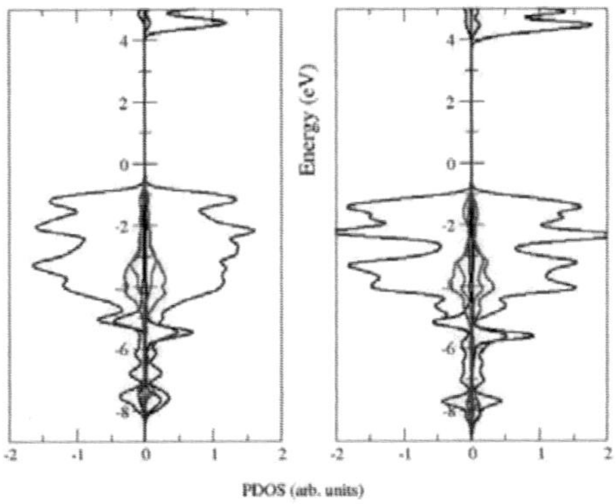

Figure 7. Partial density of states (PDOS) for the M_2Al_{18} clusters on the left and the ε-MAl_{12} clusters on the right. The pure Al clusters are shown on the left of each plot while the GaAl structures are shown on the right for each plot. The PDOS intensities for the pure Al structures are shown along the negative *x* direction while those for the Ga_2Al_{18}/$GaAl_{12}$ are along the positive *x* direction. The Fermi level is set to *y*=0. In each graph the *M-s* projection is shown in black, the *M-p* projection is shown in red, the *M-d* projection is shown in green, and the O-*p* projection is shown in blue. Adapted from Figure 3 and S7 in Ref. [28].

3.4 Heteroatom Substitution

Our adsorption studies have demonstrated shape-reactivity relationships in Keggin-based Al_{30} reactivity. More generally speaking, there is great interest in polyoxometalate chemistry of structurally and compositionally diverse clusters. Recently, our collaborators succeeded in synthesizing a new gallium(III) substituted cluster with the formula $[Ga_2Al_{18}O_8(OH)_{36}(H_2O)_{12}]^{8+}$, or Ga_2Al_{18} . The Ga_2Al_{18} cation contains a centrally-located GaO_4 tetrahedron, surrounded by octahedrally coordinated Al^{3+} atoms that arrange in trimers of the cluster. A question posed by our collaborators was whether or not the Ga substitution offered some sort of stability to the structure, as they were not able to isolate the pure Al analog, Al_2Al_{18}. To this end, we carried out DFT calculations on ε-Al_{13} and ε-$GaAl_{12}$, both of which have been isolated experimentally, and compared the electronic structure of the Keggin tridecamers to the experimental Ga_2Al_{18} structure and its hypothetical analog Al_2Al_{18}, electronic structure comparison shown in Figure 7. Based on a side-by-side comparison of the bulk atom-projected density of states that show a common electronic "fingerprint" across all four structures, we suggest that there is no unique stabilization provided by Ga substitution, and instead it may be kinetic or experimental factors that exclude the Al_2Al_{18} species. [28]

4. Future Work

While our comparison of electronic structure as a function of heteroatom substitution did not reveal unique features from tetrahedral Ga(III), questions remain regarding heteroatom substitution in molecular clusters. Other aspects of the reactivity of Keggin clusters have been shown to depend on composition. For example, work done by Casey *et al.* has studied the oxygen-isotope exchange kinetics of the Al_{13}, $GaAl_{12}$, and $GeAl_{12}$ clusters; the core oxygen atoms are inert and can be tagged with ^{17}O to allow for a comparison of all other oxygen atoms exchange rates. [30] The reactivity trend for isotopic exchange was determined to be $GeAl_{12}$ > ε-Al_{13} > $GaAl_{12}$. [8,32] Therefore, in ongoing work, we are conducting a more detailed comparison of substituted tridecamer polycations, including more detailed electronic structure and vibrational analyses.

5. Acknowledgements
We would like to acknowledge the research group of Prof. Tori Z. Forbes at the University of Iowa for their experimental collaboration. We would like to thank the University of Iowa's High Performance Computing for computing resources as well as the University of Iowa's College of Liberal Arts and Sciences, NSF award CHE-1254127, and the Center for Global and Regional Environmental Research for financial support.

References
[1] Eng PJ, Trainor TP, Brown GE, Waychunas GA, Newville M, Sutton SR, Rivers ML 2000 *Science* **288** 1029
[2] Fenter P, Cheng L, Rihs S, Machesky M, Bedzyk MJ, Sturchio NC 2000 *J Colloid Interf Sci* **225** 154
[3] Fenter P, Sturchio NC 2004 *Prog. Surf. Sci.* **77** 171
[4] Catalano JG 2010 *J. Phys. Chem. C* **114** 6624
[5] Catalano JG, Park C, Fenter P, Zhang Z 2008 *Geochim. Cosmochim. Ac.* **72** 1986
[6] Rosso KM, Yanina SV, Gorski CA, Larese-Casanova P, Scherer MM 2010 *Environ. Sci. Technol.* **44** 61
[7] Abadia M, Gonzalez-Moreno R, Sarasola A, Otero-Irurueta G, Verdini A, Floreano L, Garcia-Lekue A, Rogero C 2014 *J. Phys. Chem. C* **118** 29704
[8] Casey WH, Rustad JR, Banerjee D, Furrer G 2005 *J. Nano. Res.* **7** 377
[9] Furrer G, Phillips BL, Ulrich KU, Pothig R, Casey WH 2002 *Science* **297** 2245
[10] Abeysinghe S, Corum KW, Neff DL, Mason SE, Forbes TZ 2013 *Langmuir* **29** 14124
[11] Abeysinghe S, Unruh DK, Forbes TZ 2012 *Cryst. Growth Des.* **12** 2044
[12] Abeysinghe S, Unruh DK, Forbes TZ 2013 *Inorg Chem* **52** 5991
[13] Ilgen AG, Foster AL, Trainor T 2012 *Geochim. Cosmochim. Ac.* **94** 128
[14] Wilson SC, Lockwood PV, Ashley PM, Tighe M 2010 *Environ. Pollut.* **158** 1169
[15] Read EK, Ivancic M, Hanson P, Cade-Menun BJ, McMahon KD 2014 *Water Res.* **62** 229
[16] Mertens J, Rose J, Wehrli B, Furrer G 2016 *Water Res.* **88** 844
[17] Mertens J, Rose J, Kagi R, Chaurand P, Plotze M, Wehrli B, Furrer G 2012 *Environ. Sci. Technol.* **46** 7310
[18] Mertens J, Casentini B, Masion A, Pothig R, Wehrli B, Furrer G 2012 *Water Res.* **46** 53
[19] Mertens J 2011 *Rev. Environ. Sci. Bio.* **10** 111
[20] Brown ID 2009 *Chem. Rev.* **109** 6858
[21] Brown GE, Calas G 2011 *Cr. Geosci.* **343** 90
[22] Johansson G 1960 *Acta. Chem. Scand.* **14** 771
[23] Delley B 1990 *J. Chem. Phys.* **92** 508
[24] Delley B 2000 *J. Chem. Phys.* **113** 7756
[25] Perdew JP, Burke K, Ernzerhof M 1996 *Phys. Rev. Lett.* **77** 3865
[26] Corum KW, Mason SE 2015 *Mol. Simulat.* **41** 146
[27] Corum KW, Fairley M, Unruh DK, Payne MK, Forbes TZ, Mason SE 2015 *Inorg. Chem.* **54** 8367
[28] Fairley M, Corum KW, Johns A, Unruh DK, Basile M, de Groot J, Mason SE, Forbes TZ 2015*Chem. Comm.* **51** 12467
[29] Rustad JR 2005 *Geochim. Cosmochim. Ac.* **69** 4397
[30] Casey WH, Rustad JR 2007 *Annual Rev. of Earth and Planet. Sci.* **35** 21
[31] Wilkins RG *The study of kinetics and mechanism of reactions of transition metal complexes*; Allyn and Bacon, Inc: Boston, MA, 1974
[32] Rustad JR, Casey WH 2012 *Nat. Mater.* **11** 223

Journal of Physics: Conference Series **750** (2016) 012008 doi:10.1088/1742-6596/750/1/012008

A mean-field analysis of the simple model of evolving open systems

Takashi Shimada

Department of Applied Physics, Graduate School of Engineering, The University of Tokyo, JAPAN

E-mail: shimada@ap.t.u-tokyo.ac.jp

Fumiko Ogushi

Advanced Institute for Materials Research, Tohoku University, JAPAN

Abstract. A recently reported mechanism of letting evolving systems grow only if the interactions in it is moderately sparse is reviewed and examined. It is shown that the mean field analysis, which is known to give a good simple understanding for this transition from growing to non-growing phase but with about 30% difference in its position, is well improved by taking only mean field type information from the simulation of the original model. This supports the validity of the understanding for the mechanism of the transition, obtained from the mean field analysis. The transition stems from an essential balance of two effects: although having more interactions makes each node robust, it also increases the impact of the loss of a node.

1. Introduction

An important and universal feature of real complex systems, such as social, economic, engineering, ecological, and biological systems, is that those are evolving and open: in those systems, constituting elements are not fixed and the complexity emerges (at least persist) under successive introductions of new elements. Those systems sometimes grow, but also sometimes collapse. Therefore why and when, in general, we can have such evolving complex systems is a fundamental question.

In the previous works, we revisited this classical problem using our very simple model. It was found that the model gives either continuous growth or stagnation in system size, depending on the model's unique parameter: the average number of interactions per element m. By introducing a mean-field assumption, it was also found that this transition originates from an essential balance of two effects: although having more interactions makes each element robust, it also increases the impact of the loss of an element [1, 2]. This novel relation might be a origin of the moderately sparse network structure ($\sim$ 10 of average degree, independent of the size of the system [3]) ubiquitously found in real systems. In this paper, we verify the mean field model to figure out the origin of the difference between the previously estimated transition point $m \sim 13$ and the real transition point found in simulation, $m \sim 18.5$.

2. Transition in a simple model of evolving open systems

2.1. The simple model of evolving open systems

In the model, the entire system consists of featureless nodes, connected by directed and weighted links. What the nodes represent can be chemical, gene, animal, individual, or other species, or other diverse kind of elements. The links can be also represent various kinds of interactions or influences among them. The influence of species j on species i is denoted by the weight of the link from node j to node i, a_{ij}. Each species has only one property, "fitness", which is determined by the simple sum of interactions incoming from other species in the system:

$$f_i = \sum_{j}^{incoming} a_{ij}. \tag{1}$$

Each species can survive as long as its fitness is larger than zero, otherwise that go extinct. This extinction process represents the system's intrinsic fast dynamics.

When all the species have positive fitness, we regard that the system is in a stable (persistent) state. In such situation, nothing will happen according to the system's intrinsic dynamics. So we can wait for arbitrary long time, until a new node is introduced by some other slow process. This time scale may corresponds to the evolutionary time scale for ecosystems and biological systems, developmental/learning time scale for coupled neurons, the time scale in which a new individual joins to social communities, etc. And to focus on the most fundamental aspect of evolving systems, we give each new node m random interactions with the resident species. Note that m, the number of links we give for each new species, is the only one model parameter. The following shows the complete description of the model process.

The Evolving Open Systems (EOS) Model

(0) (Create an initial state, for example, randomly.)

(1) Calculate the fitness for each nodes: $f_i = \sum_{j}^{incoming} a_{ij}$

(2) If the fitness of the nodes are all positive, go to (3). If not, delete the species of minimum (and therefore negative) fitness and then re-evaluate the persistence of the system:
  (i) Delete the nodes
  (ii) Delete the links connecting to and from that nodes.
  (iii) Re-evaluate the extinction: go back to (1).

(3) A new node is added to the system.
  (i) Add a new node, and give m new links from or to it.
  (ii) The connecting nodes are chosen randomly from the resident nodes, with equal probability $1/N(t)$.
  (iii) The direction of the link is also determined randomly with equal probability 0.5 for each direction.
  (iv) The weights of the links are drawn from the standard normal distribution.

(4) Go back to (1) to simulate the (fast) dynamics of the new community.

2.2. The transition in growing behavior

In this model, both the introduction rule and the survival condition are neutral. Therefore it is not easy to foresee whether the system can grow under such process. Simulation results indeed show a fascinating answer: both can happen, depending on the only one parameter m, the number of interactions per species. The system can grow to infinitely large if the number of the interactions per species is in a moderate range ($5 \leq m \leq 18$), and, if not, it stays in a finite size (Table 1).

3. A mean-field model of the EOS model and its verification

3.1. A mean-field model

The structure of the emergent network is found to have no strong deviation from that of Erdős-Rényi random graph with an average number of links $\langle m \rangle \approx m$. Taking such a correlation-less structure as a support, the following mean field approximation was made to better understand the mechanism of the transition.

In our mean field treatment, we describe the system by the distribution function of the fitnesses of the species in it. The change in the fitness of each species in the original model arises from the addition or deletion of an incoming link. Although each such event changes both the network structure and the fitnesses, we here treat only the fitness distribution. Under the mean field treatment, such link change events act as a diffusion process to the fitness distribution function. More precisely, the diffusion is neutral for link addition events but that involves negative drift for link deletion events. Therefore, the process of inclusion of new species is represented by the diffusion process with negative drift and an addition of initial fitness distribution (positive half of a Gauss distribution). The extinction process of the species with non-positive fitness is modeled by making a cut of the fitness distribution function under zero after applying the random walk. The equilibrium fitness distribution is obtained as a fixed distribution against these processes. And once we get the equilibrium distribution numerically, the area in the negative fitness part of the distribution after applying the random walk process gives the probability E of a resident species going extinct during one link addition/deletion event. The average number of nodes that go extinct directly because of the introduction of the new node is $mE/2$. And these extinctions may also trigger further extinctions. If we approximate the graph by a loop-less random network, the expectation value of the total number of extinctions per addition of one node into a system in which all species have m interactions can be calculated as an infinite geometric series as,

$$N_E = \sum_n \left(\frac{mE}{2}\right)^n = \frac{mE}{2 - mE}. \tag{2}$$

Because $N_E = 1$ means that the number of extinctions is equal to the number of additions in the long-term average, $mE = 1$ corresponds to the transition point. This criteria give us the correct transition behavior, i. e. finite phase to divergent phase as m increases. However, the predicted transition point from this mean field model, $m_c \sim 13$, is smaller than that obtained from the simulation by about 30%.

Table 1. The phase diagram of the EOS model. The transition at between $m = 18$ and 19 The transition at between $m = 4$ and 5 simply stems from the sparseness of the network (i.e. percolation) and hence is not re-examined in this study.

m :	≤ 4	$5 \sim 18$	≥ 19
$N(t)$:	finite	diverging	finite

Journal of Physics: Conference Series **750** (2016) 012008 doi:10.1088/1742-6596/750/1/012008

3.2. Verification of the mean-field model

The possible cause of the difference between the real transition point and that of mean field analysis are two: neglecting any structural correlation and taking the tree structure for the calculation of N_E, and the error in the estimation of the fitness function and E. To verify which of these two contributions is dominant, we here evaluate the extinction probability directly from the simulation. The extinction probability in an emergent network can be calculated as

$$\begin{aligned} E_i &= \frac{1}{N}\sum_i^N \Phi(-f_i), \\ E_e &= \frac{\#_{link}(a_{ij} > f_i)}{\#_{link}}, \end{aligned} \tag{3}$$

where E_i and E_e are the extinction probability during the link addition and deletion events. f_i and a_{ij} are the fitness of the ith node and the weight of the link from jth node to ith node in the emergent network, and

$$\Phi(x) = \frac{1}{\sqrt{2\pi}}\int_{-\infty}^{x} e^{-\frac{t^2}{2}}\, dt \tag{4}$$

is the cumulative distribution of the standard normal distribution. Substituting E by these "empirically" obtained extinction probabilities, the same mean-field treatment gives the average number of extinctions as

$$\langle N_E\rangle = \frac{mE_i}{2}\left[1 + \frac{\langle m\rangle E_e}{2} + \left(\frac{\langle m\rangle E_e}{2}\right)^2 + \cdots\right] = \frac{mE_i}{2 - \langle m\rangle E_e}, \tag{5}$$

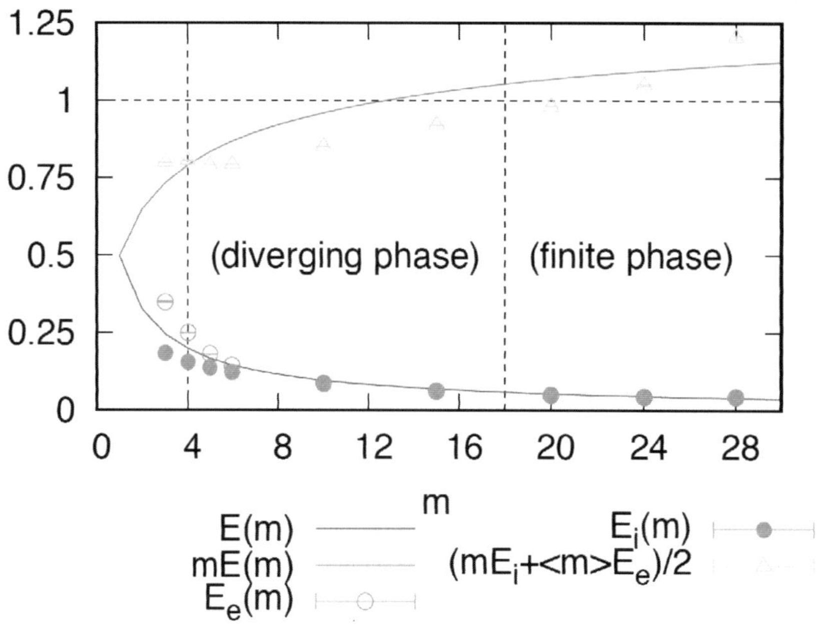

Figure 1. Estimation of the transition point, using the semi-analytical mean field theory, ($mE = 1$) [2], and the mean field treatment with empirically obtained extinction probabilities (Eq. (5)).

where $\langle m \rangle$ is the average degree of the emergent network. Therefore this empirical test predicts the transition at

$$\frac{mE_i + \langle m \rangle E_e}{2} = 1. \tag{6}$$

As shown in Figure 1, while the extinction probability in the original self-consistent mean field theory is over estimated, the empirically obtained average extinction probability predicts the transition point almost perfectly, which means that the correction due to the correlation among the interaction coefficients, the loop structure (i.e. non-zero clustering coefficient), and so on of the emergent network is very small. This result suggests the validity of basic understanding of the transition mechanism obtained from the mean field picture: having more links makes each node more robust, but the net impact of loosing a node increases hence in total the robustness of the entire system decreases.

4. Conclusions

We verified the mean field approximation for the simple EOS model, using the information of emergent networks in the original model. The good agreement in the transition point, without taking any structural information into account, support the validity of the mean field treatment. This confirms that the transition essentially stems from an essential balance of two effects: although having more interactions makes each node robust, it also increases the impact of the loss of a node.

Acknowledgement

This work was supported by JSPS KAKENHI Grant Number 15K05202 to TS and CREST, JST.

[1] T. Shimada, in Springer monograph *Mathematical Approaches to Biological Systems: networks, Oscillations and Collective Motions* (Springer, 2015) p. 95-117.

[2] T. Shimada "A universal transition in the robustness of evolving open systems", Scientific Reports, Vol. 4, 4082 (2014).

[3] R. Albert and A.-L. Barabási "Statistical mechanics of complex networks", Reviews of Modern Physics, Vol. 74, p. 47-97 (2001).

Comparing atomistic and coarse–grained simulations of P3HT

Jonathan Gross, Momchil Ivanov and Wolfhard Janke

Institut für Theoretische Physik and Centre for Theoretical Sciences (NTZ), Universität Leipzig, Postfach 100 920, D-04009 Leipzig, Germany

E-mail: gross@itp.uni-leipzig.de, momchil@itp.uni-leipzig.de, janke@itp.uni-leipzig.de

Abstract. Poly(3–hexylthiophene) (P3HT) is a key material used in organic photovoltaics. In this study we assess the validity of two coarse–grained models of P3HT by comparing coarse–grained Monte Carlo simulations to fully atomistic molecular dynamics simulations. Structural properties of single polymer chains of short to medium lengths are compared between the three representations.

1. Introduction

The interest in Poly(3–hexylthiophene) (P3HT) comes from its application in organic photovoltaics due to its electronic and optical properties [1, 2]. For a better understanding of the polymer in interaction with other materials several experimental and theoretical studies have been published in recent years [3, 4, 5, 6]. It is a very common simulational approach to use coarse–grained models to study a wide range of properties of a given system. With this strategy it is much faster to sample the phase space compared to an all–atom study, but this comes at the cost of losing some detail. In a previous work [7] we investigated the adsorption behavior of P3HT on reconstructed gold surfaces using experimental and simulational approaches (see Fig. 1 for exemplary structures). In the current work we compare two different types of coarse–grained models using Monte Carlo (MC) simulations with all–atom molecular dynamics (MD) simulations. We want to determine which one of the two coarse–grained models matches the all–atom simulations best.

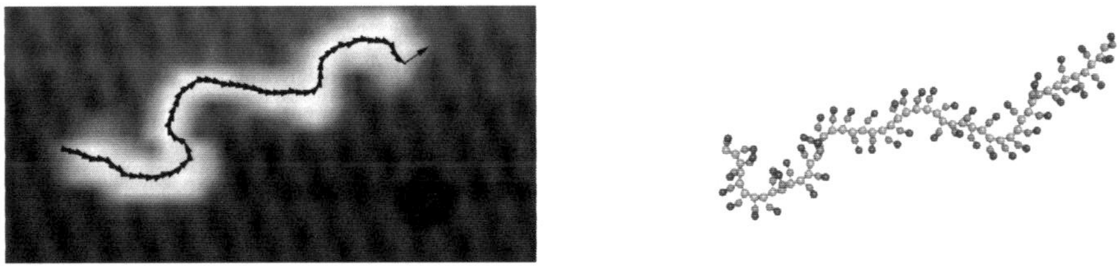

Figure 1. Scanning tunneling microscopy image of a P3HT 50–mer on Au(001) (left) and elongated 50–mer conformation on a surface obtained by Monte Carlo simulations (right).

Figure 2. Schematic coarse–graining idea (after Ref. [3]) (left) and resulting homopolymer model (right).

2. Models

In this section we introduce the two coarse–grained (CG) models and the parameterization of the all–atom MD simulations. The first CG representation (Model A) we investigated in this work is a simple homopolymer model where every unit of P3HT has been coarse–grained into a single CG bead at the center of mass of the corresponding monomers of the atomistic chain. The interactions between the CG particles of Model A are given by a harmonic potential to model the bonds $U_{\text{bond}}(l) = \frac{1}{2}k_b \cdot (l - l_0)^2$, a harmonic term for the bending energy $U_{\text{bending}}(\Theta) = \frac{1}{2}k_\Theta \cdot (\Theta - \Theta_0)^2$, a dihedral angle potential of the following form $U_{\text{torsion}}(\Phi) = \frac{1}{2}\sum_{n=1}^{3} V_n \left[1 + (-1)^{n-1}\cos(n\Phi)\right]$, and finally Lennard-Jones like interactions for all non–bonded particles $U_{\text{non-bonded}}(r) = 4\epsilon\left[\left(\frac{\sigma}{r}\right)^{12} - \left(\frac{\sigma}{r}\right)^{6}\right]$. To obtain the 9 parameters for these potentials, atomistic MD simulations have been run and the iterative Boltzmann inversion method [8, 9] was applied. For details and the actual parameters please refer to [3]. This model was originally developed to describe the interaction of P3HT with a fullerene derivate phenyl-C_{61}-butyric acid methyl ester (PCBM) in melts as they occur in aforementioned solar cells. A schematic representation of the coarse–graining ansatz and the resulting homopolymer bead–spring model are depicted in Fig. 2.

The second representation (Model B) allows for some more detail since it includes information on the side–chains (see Fig. 3 on the left). Here we distinguish three coarse–grained particles. P1 are the backbone particles with their origin in the center of mass of the thiophene ring. The side–chains in this CG model are represented by two beads P2 and P3, each including three methyl groups. A sketch of the structure can be found in Fig. 3 on the right. As with Model A the parameters of the CG potentials for this model were determined by MD simulations and calculated using the iterative Boltzmann inversion method [4]. The resulting potential for the bonds between particles $U_{\text{bond}}(l) = \sum_{i=2}^{n_b} c_i (l - l_0)^i$, has a total of 26 parameters for the three

Figure 3. Atomistic structure of a P3HT 3–mer in bead-stick representation (left) and a sketch of the three beads per monomer coarse–grained Model B (right).

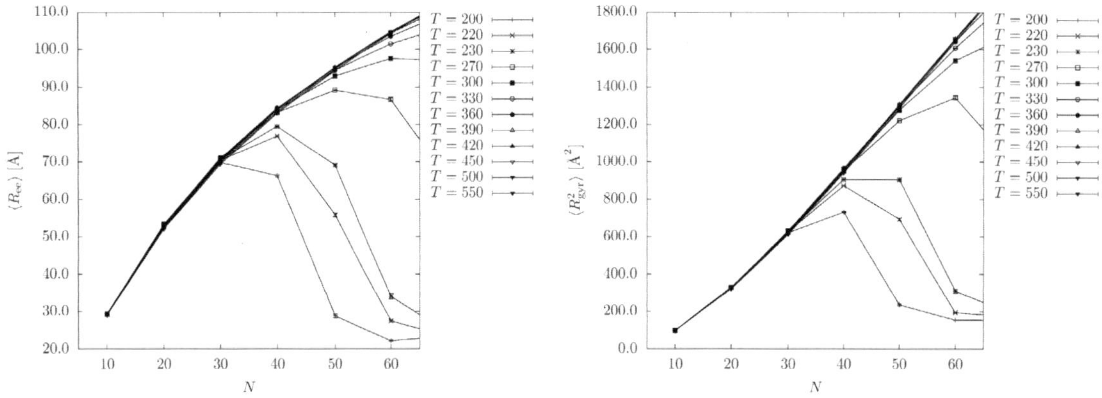

Figure 4. Model A: End–to–end distance (left) and squared radius of gyration (right) for lengths $N = 10$ to 60 at temperatures in the range from $T = 200\,\mathrm{K}$ to $550\,\mathrm{K}$.

types of possible bonds (P1-P1, P1-P2, and P2-P3). Angles between two bonds contribute to the total energy via $U_{\text{bending}}(\Theta) = \sum_{i=0}^{n_a} c_i \left(\Theta - \Theta_0\right)^i$, with a total of 42 parameters. The dihedral interaction is described by $U_{\text{torsion}}(\Phi) = \sum_{i=0}^{n_t} \cos^i(\Phi)$ with 24 parameters. Finally, the last potential to model bonded interactions is a polynomial with 5 parameters describing the improper dihedral for P1-P2-P1-P1. Additionally, the original paper for this model includes 4 pages of tabulated values which describe the non–bonded interaction, which include Lennard–Jones and Coulomb interactions. This leads to a total number of 97 parameters for this coarse–grained model, not counting the non–bonded table. It is rather unusual to have such a large number of parameters, since the idea of coarse-graining is the reduction of degrees of freedom.

To simulate the atomistic model we employ the all–atom variant of the Optimized Potentials for Liquid Simulations (OPLS-AA). This force field contains four contributions, the bonds using $U_{\text{bond}}(r) = \frac{1}{2}k_b \left(r - r_0\right)^2$, the angles with $U_{\text{bending}}(\Theta) = \frac{1}{2}k_\Theta \left(\Theta - \Theta_0\right)^2$, the third contribution has two terms describing the proper and improper dihedrals, $U_{\text{torsion}}(\Phi) = \sum_{i=0}^{5} k_i \cos\left(\Phi - \pi\right)$ and $U_{\text{improper}}(\Phi) = k\left[1 + d\cos\left(n\Phi\right)\right]$, and lastly the non-bonded contribution, which includes Lennard-Jones and Coulomb interactions. An exemplary structure of a 3–mer is shown in Fig. 3 on the left.

3. Geometrical comparison

For models A and B we performed replica-exchange Monte Carlo simulations at 12 temperatures in the range from $T = 200\,\mathrm{K}$ to $550\,\mathrm{K}$ for different numbers of monomers N ranging from 10 to 60. The quantities we want to focus on in this study are two basic geometric properties that are also accessible in experiment and thus allow for comparison with our simulational results. Firstly we look at the end–to–end distance $R_{\text{ee}} = |\vec{r}_N - \vec{r}_1|$ and secondly the squared radius of gyration $R^2_{\text{gyr}} = R^2_{xx} + R^2_{yy} + R^2_{zz}$, with $R^2_{ij} = \frac{1}{N}\sum_{n=1}^{N}\left(r_i^{(n)} - \overline{r}_i\right)\left(r_j^{(n)} - \overline{r}_j\right)$ where N is the number of monomers, $r_i^{(n)}$ $(i = x, y, z)$ is the i-th Cartesian coordinate of the n-th monomer and $\overline{r}_i = \sum_{n=1}^{N} r_i^{(n)}/N$.

Figure 4 shows the end–to–end distance (left) and the squared radius of gyration (right) for Model A. Both quantities show the same behavior and allow for the same conclusions. All chains are elongated for temperatures above 270 K and chains that are longer that 30 monomers start collapsing at temperatures below 270 K. Noticeable is that at room temperature (300 K),

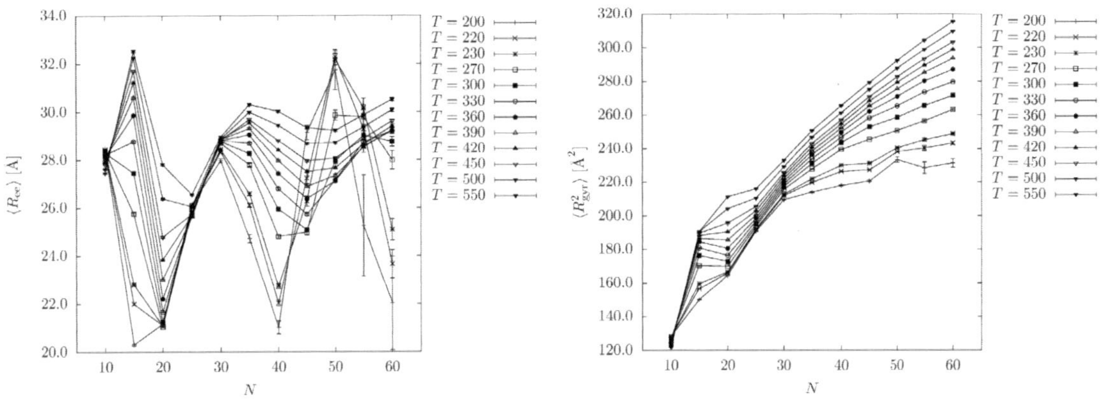

Figure 5. Model B: End–to–end distance (left) and squared radius of gyration (right) for lengths $N = 10$ to 60 at temperatures in the range from $T = 200\,\text{K}$ to 550 K.

all chains are elongated. A fit to the radius of gyration data at that temperature results in a $R_{\text{gyr}} \sim N^{0.74}$ scaling.

We show the same quantities for Model B in Fig. 5, but this time we see a completely different behavior. The end–to–end distance is almost the same for all chain lengths and rather short, suggesting a dense or collapsed structure. The temperature seems to have virtually no influence on the end–to–end distance. Also for the radius of gyration the temperature only plays a minor role. The values of the radius of gyration also suggest much more closely packed structures, when compared to those of Model A. We come to the conclusion that in this model the chains at all temperatures are in a globule state, which agrees with the scaling for the radius of gyration that we again measure at room temperature to be $R_{\text{gyr}} \sim N^{0.24}$.

For the atomistic model we employed molecular dynamics simulations using the GROMACS software package. We used the Velocity-Verlet integrator in combination with the massive Andersen thermostat. We decided to use a stochastic thermostat, since we are only interested in structures at certain temperatures and can neglect dynamics at this point. Simulations were run at $T = 300\,\text{K}$ using a time step of 1 fs, which is of the order of the bond fluctuations. For

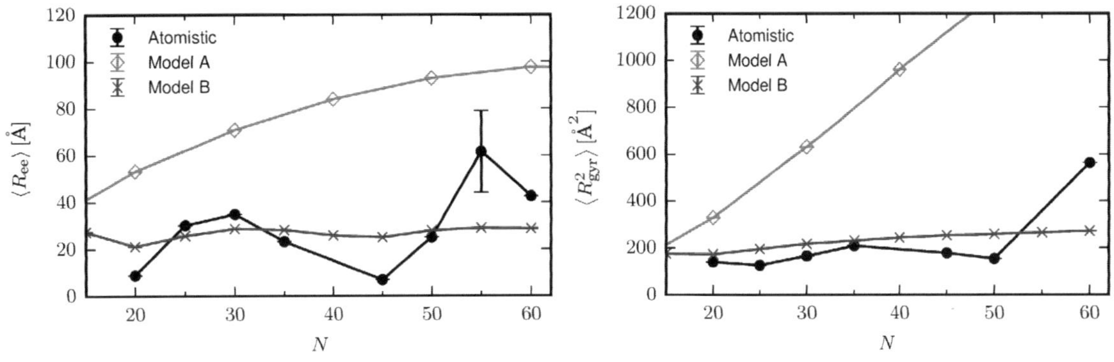

Figure 6. End-to-end distance (left) and squared radius of gyration (right) for the coarse–grained models A and B and the atomistic model at $T = 300\,\text{K}$. Errors for the atomistic model are of the order of the exemplary one shown for $N = 55$. For models A and B error bars are of the size of the symbols.

each chain length a trajectory of 100 ns was recorded. The results for the end–to–end distance and the squared radius of gyration are shown together with the respective results from Model A and B in Fig. 6. Even though the fluctuations in the MD results are quite large, especially for long chains, due to the slow convergence of MD simulations in general, the agreement of the results with those from Model B are indisputable. The polymer is in a globule state for all chain lengths at room temperature.

To examine the source of the difference in the result we look at the persistence length as a final quantity. The persistence length l_p is here defined as the length where the average $\cos\Theta(s)$ reaches the value $1/e$, with $l_b \approx 4\,\text{Å}$, $s = |i-j|$ and

$$\langle \cos\Theta(s)\rangle = \langle \vec{a_i}\cdot\vec{a_j}\rangle / \langle a_i^2\rangle = \exp(-sl_b/l_p), s \to \infty.$$

This is also the standard calculation provided by the GROMACS package. While Model A exhibits persistence lengths between 30 Å and 40 Å, Model B and the atomistic simulations agree on a persistence length of only approximately 15 Å. This leads us to the conclusion that the parameterization for Model A overestimates the stiffness of a single polymer and that Model B is able to reproduce the stiffness behavior more appropriately.

Finally, we want to comment on the computing time involved in the above simulations. The times we give here are in units of core hours, although the actual simulations are, of course, run in parallel. This helps to put the computational effort of the different approaches in perspective. For the one–particle bead–spring polymer (Model A) we ran parallel–tempering simulations at 12 temperatures for 15 chain lengths with 10^8 Monte Carlo sweeps each. The total run time was 7800 hours which is approximately 325 days. Simulations of Model B at 12 temperatures for 11 chain lengths with 10^7 sweeps each took about 2000 hours corresponding to about 83 days. The atomistic simulations of 100 ns trajectories for 9 chain lengths and only one temperature took about 11000 hours which equals to about 450 days.

4. Summary

Our simulations confirm the suitability of Model B to realistically reproduce geometrical observables of single isolated P3HT polymer chains when compared to all–atom molecular dynamics simulations. Since the original parameterization of Model A was to replicate the behavior of P3HT with fullerene in melts, we find it not suitable for our single polymer studies. The persistence length and thus the stiffness of the polymer is clearly overestimated in Model A. Furthermore our comparison proves the striking advantage of coarse–grained Monte Carlo simulations compared to MD simulations in terms of efficiency and statistics.

Acknowledgments

This work has been supported by the DFG (German Research Foundation) under Grant No. SFB/TRR 102 (Project No. B04). All simulations were run at the compute cluster within the Institute for Theoretical Physics of the University of Leipzig.

References

[1] Xu B and Holdcroft S 1993 *Macromolecules* **26** 4457
[2] Boa Z, Dodabalapur A and Lovinger A J 1996 *Appl. Phys. Lett.* **69** 4108
[3] Lee C K, Pao C W and Chu C W 2011 *Energy Environ. Sci.* **4** 4124
[4] Huang D M, Do K and Moule A J 2010 *J. Chem. Theory Comput.* **6** 526
[5] Liu Y F, Krug K and Lee Y L 2013 *Nanoscale* **5** 7936
[6] Förster S and Widdra W 2014 *J. Chem. Phys.* **141** 054713
[7] Förster S, Kohl E, Ivanov M, Gross J, Widdra W and Janke W 2014 *J. Chem. Phys.* **141** 164701
[8] Reith D, Pütz M and Müller-Plathe F J 2003 *J. Comput. Chem.* **24** 1624
[9] Reith D and Faller R 2003 *Macromolecules* **36** 5406

Knots in finite memory walks

Eric Horwath[1], **Nathan Clisby**[2], **Peter Virnau**[1]

[1] Institut für Physik, Johannes Gutenberg-Universität Mainz, Staudinger Weg 7, 55099 Mainz, Germany

[2] School of Mathematics and Statistics, The University of Melbourne, Victoria 3010, Australia

E-mail: nclisby@unimelb.edu.au, virnau@uni-mainz.de

Abstract. We investigate the occurrence and size of knots in a continuum polymer model with finite memory via Monte Carlo simulations. Excluded volume interactions are local and extend only to a fixed number of successive beads along the chain, ensuring that at short length scales the excluded volume effect dominates, while at longer length scales the polymer behaves like a random walk. As such, this model may be useful for understanding the behavior of polymers in a melt or semi-dilute solution, where exactly the same crossover is believed to occur. In particular, finite memory walks allow us to investigate the role of local interactions in the transition from highly knotted ideal polymers to almost unknotted self-avoiding polymers. Even though knotting decreases substantially when a few next-nearest neighbor interactions are considered, we find that the knotting probability of a polymer chain of modest length of 500 steps only decays slowly as a function of the range of the excluded volume interaction. In this context, we also find evidence that for length scales up to the interaction length the knotting behavior of the finite memory walk resembles that of a self-avoiding walk (effectively suppressing small knots), while for larger length scales it resembles that of a random walk.

1. Introduction

Knots are known since the dawn of mankind for their utility and aesthetics, but have only been described in mathematical terms rather recently. Arguably, the foundations of modern knot theory have been laid by two Scottish physicists in the 1870s and 1880s: Lord Kelvin, who devised an atomic theory based on the idea of vortex atoms made from ether [1] and Peter Tait, who started to tabulate knots in an attempt to account for all vortex atoms [2].

A variant of Tait's classification scheme [3] is still used today and categorizes knots according to the minimum number of crossings of a closed loop in a projection onto a plane. Apart from the so-called unknot, an unknotted circle, there is one knot with three, one with four (see figure 1), two knots with five crossings, three knots with six and so on. This number increases, however, exponentially and there are, e.g., already more than one million distinct knots with more than 16 crossings [4]. It is therefore not surprising that there is still no algorithm which is able to distinguish between all knots. As knots are only mathematically defined in closed loops, the termini of open chains are typically connected in a prescribed way before knot detection takes place [5; 6].

In the 1960s interest in knots among natural scientists was revived by the Frisch-Wasserman-Delbrück conjecture [7; 8]. This conjecture essentially states that all polymers (just like macroscopic strings) will eventually become knotted if they are just long enough. This was proved in the case of self-avoiding walks on the lattice by Sumners and Whittington [9]. The

Figure 1. From left to right: Unknot, Trefoil (3_1) and Figure-Eight-Knot (4_1). The index was introduced by Alexander and Biggs to distinguish between knots with the same crossing number.

theoretical work of the 1960s eventually led to the discovery and creation of knots in DNA [10–15], proteins [16–24] and polymers [25; 26], and a multitude of simulation work e.g. [27–45] to better understand their properties. Intriguingly, two of the basic model classes have already been addressed in the first two simulations, respectively simple random walks and self-avoiding walks in 1974 and 1975 [27; 28].

Simple random walks (also known as ideal chains), either on the lattice or in the continuum, are the simplest model of a polymer as there are no interactions at all between monomers. In this case, although the short distance behavior may depend on details such as the lattice, the large distance behavior converges to Brownian motion, and all such models belong to the same universality class.

Self-avoiding walks are a more realistic model of polymers which incorporate the excluded volume interaction. On the lattice they are defined as the subset of random walks that are strictly self-avoiding, i.e. no two monomers can occupy the same lattice site. It is widely believed that dilute solutions of polymers in a good solvent, self-avoiding walks on a lattice, and a wide variety of related models on the lattice and in the continuum with sufficiently strong excluded volume interaction, are all in the same universality class.

One key observable which characterises the behavior of a polymer model is the mean-square end-to-end distance, and the dependence of this observable on the number of monomers, N, is given by

$$\langle R_{\mathrm{e}}^2 \rangle = D N^{2\nu}, \tag{1}$$

where D is a model- and lattice-dependent constant, and ν is a universal critical exponent which is identical for all models in the same universality class. For simple random walks, $\nu = 1/2$, as expected for Brownian motion, while for three-dimensional self-avoiding walks $\nu = 0.587597(7)$ [46].

Random walks [27] knot readily as it is easy to create loops and knots on a very local scale. Typically, six to seven segments suffice for a simple trefoil [35]. This behavior changes drastically if repulsive excluded volume interactions are considered [28], which model polymers in good solvent conditions. Local entanglements disappear and the probability of observing knots becomes almost zero for chains of comparable size. Even though these knots are localized as well [37; 39], they are typically larger than knots in random walks. The transition from highly knotted ideal chains to almost unknotted self-avoiding polymers was first investigated in 1991 by Koniaris and Muthukumar [30; 47] who determined knotting probabilities as a function of chain thickness.

However, isolated polymers are not the only system of interest, and in particular polymer melts and semi-dilute polymer solutions are of great theoretical, experimental, and industrial interest. Polymer melts consist of a dense system of self-avoiding polymer fragments that nonetheless behave like simple random walks at large length scales (i.e. $\nu = 1/2$), even though corrections to scaling exist even for large chains [48]. Similarly, the blob model of semi-dilute polymers suggests that they behave like self-avoiding walks within blobs, but the blobs themselves for a particular chain diffuse in accordance with Brownian motion. The interplay between these two length scales determines the overall exponent ν in this case.

One model which performs such an interpolation between self-avoiding walks at short lengths, and simple random walks at greater lengths, is that of finite memory walks, where excluded volume interactions still occur, but only between monomers separated within a fixed cut-off distance r along the chain.

These walks have been considered on various lattices, and have been useful in proving upper bounds on the connective constant [49–51]. This is because self-avoiding walks are a strict subset of finite memory walks for any cut-off r, and the number of finite memory walks can be exactly calculated for small values of r. More recently, the case with $r = 2$ on the lattice, so-called non-backtracking random walks, have been proved to converge to Brownian motion [52].

It is perfectly natural to extend this idea to polymer models with more complicated excluded volume interactions, either on- or off-lattice. In this work we consider an off-lattice variant of finite memory walks, and investigate the effect of the internal length scale r on the knotting behavior of the polymer.

2. Model and Algorithms

2.1. A continuum finite-memory walk

We investigate a continuous variant of a finite-memory walk [49–51]. Adjacent beads along the chain are connected with fixed bonds of length 1σ. Excluded volume interactions are defined via a repulsive Weeks-Chandler-Anderson (WCA) potential:

$$U_{\mathrm{WCA}}(d) = \begin{cases} 4\epsilon\left[\left(\frac{\sigma}{d}\right)^{12} - \left(\frac{\sigma}{d}\right)^{6} + \frac{1}{4}\right], & \text{if } d < d_{\min} = 2^{1/6}\sigma \\ 0\,, & \text{else} \end{cases} \tag{2}$$

Excluded volume interactions are only considered between beads of a distance along the chain of up to r bonds. $r = 0$ refers to a random walk with no excluded volume interactions, and in our case $r = 1$ also corresponds to the pure random walk because the bond length is fixed. For $r = 2$ next-nearest neighbors interact, and for a polymer with N monomers $r = N - 1$ corresponds to a fully self-avoiding polymer.

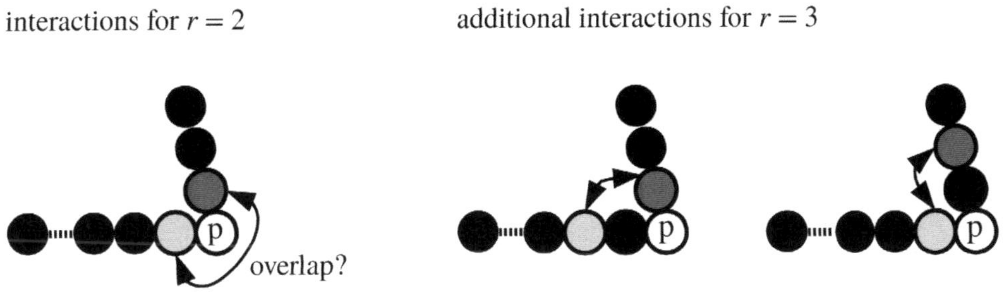

Figure 2. For $r = 3$, three interactions need to be accounted for in the determination of energy change ΔE in the Metropolis criterion.

2.2. Implementation of the pivot algorithm

Configurations are sampled with the fastest known method for polymers in a good solvent, a Markov chain Monte Carlo approach called the pivot algorithm [53; 54]. As computer time is dominated by the determination of knot type we use the standard implementation of the pivot algorithm [54] rather than exploiting recent implementations which are asymptotically faster [46; 55; 56]. Typically, a pivot center is chosen randomly and one arm of the chain rotated around the center by an arbitrary angle. For a fully self-avoiding polymer, all interactions between monomers of the rotated arm with the other arm need to be accounted for in the Metropolis criterion. Interactions between beads belonging to the same arm do not change and need not be considered. Likewise, for random walks ($r = 0, 1$) all moves are accepted automatically as there is no interaction energy and thus no change in energy. For next-nearest neighbor interactions ($r = 2$), we only determine the change in energy between the bead succeeding the pivot center on the rotated arm with the bead preceding the pivot center. For arbitrary r, we determine interactions of the first r beads with their interaction partners on the other arm including the pivot center (for $r = 3$, see figure 2).

2.3. Knot detection

In this manuscript, we apply a variant of the Alexander polynomial [57], as detailed in reference [5], which allows us to distinguish (though not unambiguously) between simple knots. As knots are only mathematically well-defined in closed curves [5; 6], the ends of linear chains need to be connected in a well-defined manner first, before the knot type is determined. In this work, we apply a closure in which two lines emerge from the termini along the respective connection lines from the center of mass of the chain to the termini [19]. Then, the two lines are connected far away from the chain. Even though knots can in principle be created by the specific closure, knotting probabilities vary only little when different closures are applied [5; 39]. The size of the knot is determined by deleting monomers first from one side until the Alexander polynomial changes, which fixes one of the ends of the knotted subchain as the last monomer before the change occurred. Monomers are then successively deleted from the remaining end of the subchain until the Alexander polynomial changes again. The number of monomers in the knotted subchain provide a measure of the extent of the knot.

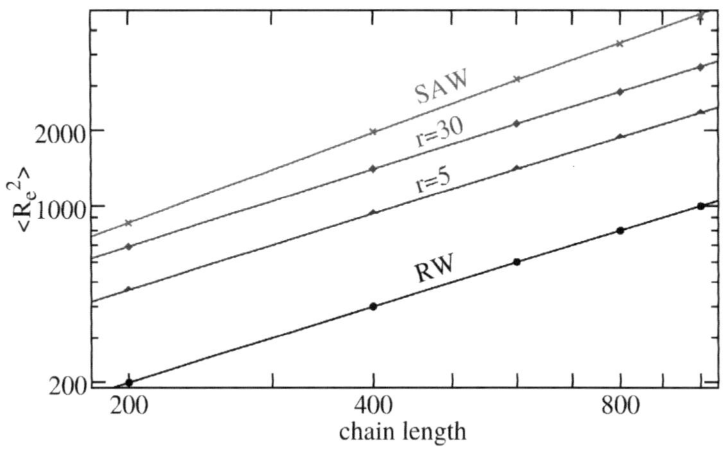

Figure 3. Mean square end-to-end distance $\langle R_e^2 \rangle$ as a function of chain length for random walk ($r = 0$), finite-memory walk ($r = 5$, $r = 30$) and self-avoiding polymer.

3. Results

3.1. Scaling

First, we would like to verify that our finite-memory walks indeed scale like random walks in the limit of long chains. From fitting the data in figure 3 we obtain scaling exponents of $\nu = 0.50$ for $r = 5$ and $\nu = 0.51$ for $r = 30$. Interestingly, our results for rather modest chain sizes are already compatible with the critical exponents for random walks ($\nu = 0.50$) and can clearly be distinguished from the Flory exponent for self-avoiding polymers $\nu = 0.587597(7)$ [46]. (For our rather modest chain sizes we have obtained $\nu \approx 0.59$.) Nevertheless, the mean square end-to-end distance of the finite memory walk is still significantly larger than the corresponding value of a random walk and resembles the result obtained for our fully self-avoiding polymer, which leads to an increase in the non-universal amplitude in the scaling law of equation 1.

3.2. Knot probability

As shown in figure 4, the probability of observing a knot quickly decreases to almost half the value observed for a random walk ($r = 0, 1$) when a few neighboring excluded volume interactions are considered. However, the knotting probability only decreases gradually afterwards and it takes $r = 30$ to decrease the probability to about 13%. Likewise, full excluded volume interactions are required to obtain the knotting behavior of a self-avoiding polymer, which is almost always unknotted ($P(\text{knot}) = 0.0014$). For comparison we have also plotted the probability for observing a trefoil knot (3_1). While for the random walk the knots predominantly have larger crossing numbers, we observe that the complexity of knots decreases with r, until for $r = 30$ almost all knots are simple trefoil knots.

3.3. Distribution of knot sizes

Figure 5 shows the distribution of knot sizes for configurations, which contain a simple trefoil knot. As indicated in the introduction the most likely size of a trefoil knot in a random walk is only around six or seven segments [35] as it is very easy to create loops on a local scale if

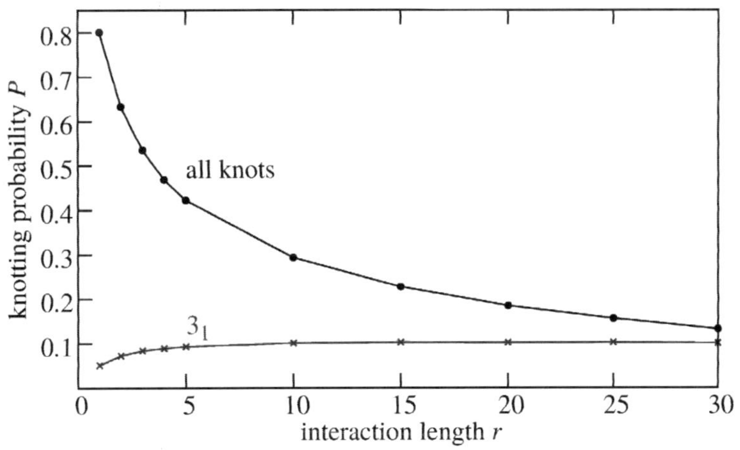

Figure 4. Probability of observing any knot for a finite-memory walk of size 500 and a trefoil knot as a function of interaction range r. $r = 0$ or 1 corresponds to a random walk. The probability P of observing a knot in a self-avoiding polymer is almost zero ($P(\text{knot}) = 0.0014$). The lines are guides to the eye. Errors are smaller than symbol sizes.

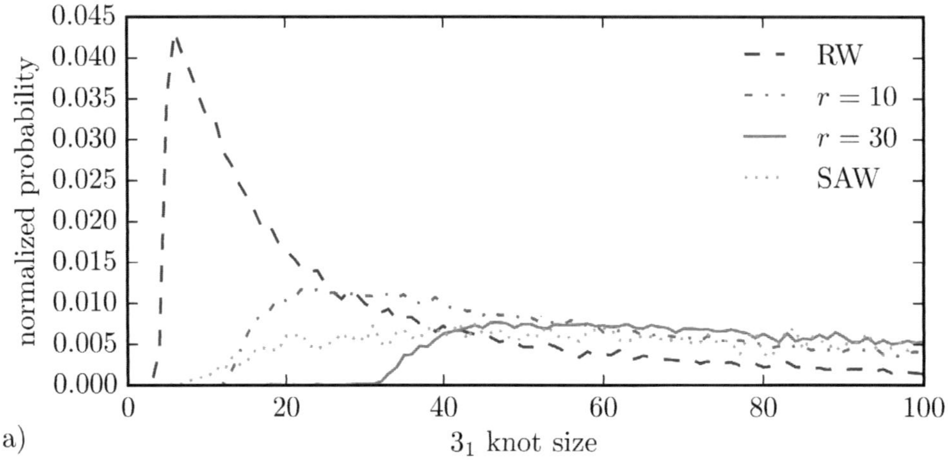

a)

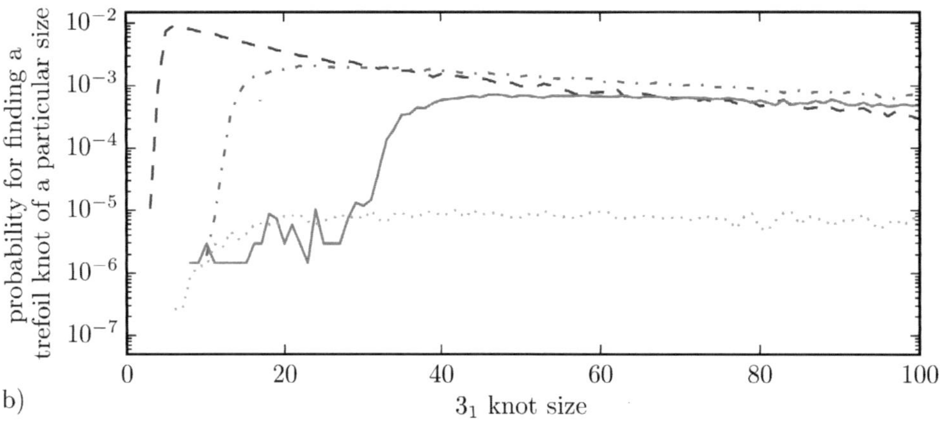

b)

Figure 5. a) Size distributions of trefoil knots for a random walk ($r = 0$), finite memory walks ($r = 10$ and $r = 30$) and a fully self-avoiding polymer ($r = 499$). In all cases, the chain consists of 500 monomers. All curves are normalized to one. b) Same as above but with a semi-logarithmic scale. In addition, all histogram entries have been divided by the total number of walks instead of the total number of trefoil (3_1) knots.

beads do not repel each other. For a fully self-avoiding polymer, the local stiffness emerging from the excluded volume interactions shifts the most likely size of a trefoil knot to larger values. Finite memory walks on the other hand exhibit a rather unexpected and peculiar behavior. In figure 5a, there are almost no knots of size up to r, while for larger knot sizes the behavior resembles that of a random walk. This feature becomes even more pronounced when we divide our histogram by the total number of walks (with fixed r) (figure 5b). Then the probability on the ordinate corresponds to the probability of observing a knot of a particular size. In this representation the knotting probability for $r = 30$ follows closely that of a self-avoiding walk up to sizes of 30 and then quickly emulates the knotting behavior of random walks for larger sizes.

4. Discussion and conclusions

In this work we study the influence of local excluded volume interactions on knottedness in polymer chains by investigating the occurrence and size of knots in continuous finite memory walks. This allows us to study the transition from highly knotted random walks to almost unknotted self-avoiding walks as a function of the extent of local interactions. Local excluded volume interactions stiffen the chain and reduce knotting substantially even though the chain still behaves like an ideal chain from the point of view of scaling. However, after the strong initial decay in the probability of finding knots as a function of the extent of local interactions, the decay flattens and it takes self-avoidance in the whole chain to obtain the knotting probability of the fully self-avoiding polymer. The strong localization of entanglements and knots observed for random walks is softened and shifted to larger lengths. We find evidence that for length scales up to interaction range r the finite memory walk behaves like a self-avoiding walk (prohibiting knots of sizes smaller than r), while for larger length scales the knotting behavior resembles that of a random walk. Finite-memory walks offer the possibility of tuning knotting properties of polymer chains while retaining the scaling properties of random walks (in contrast to self-avoiding walks with reduced excluded volume throughout the chain). They are thus interesting candidates for the structural description of single chains in a polymer melt or a semi-dilute polymer solution, which will be investigated in future work.

Acknowledgments

This work builds on an idea, which has emerged from discussions between Nathan Clisby and Peter Virnau at the 30th CSP workshop. This collaboration effort would not exist without the workshop. NC wishes to thank the Australian Research Council for supporting this work under the Future Fellowship scheme (project number FT130100972).

References

[1] Thompson W 1867 *Philos. Mag.* **34** 15–24
[2] Tait P G 1898 *Scientific Papers* **1** 273–347
[3] Alexander J W and Briggs G 1926 *Annals of Mathematics* 562–86
[4] Hoste J, Thistlethwaite M and Weeks J 1998 *Math. Intell.* **20** 33–48
[5] Virnau P 2010 *Physics Procedia* **6** 117–125
[6] Tubiana L, Orlandini E and Micheletti C 2011 *Progr. Theor. Exp. Phys.* 192–204
[7] Frisch H L and Wasserman E 1961 *J. Am. Chem. Soc.* **83** 3789
[8] Delbrück M 1962 *Proc. Symp. Appl. Math.* **14** 55–63
[9] Sumners D W and Whittington S G 1988 *J. Phys. A: Math. Gen.* **21** 1689
[10] Macgregor H C and Vlad M 1972 *Chromosoma* **39** 205
[11] Liu L F, Depew R E and Wang J C 1976 *J. Mol. Biol.* **106** 439–52
[12] Dean F B, Stasiak A, Koller T and Cozzarelli N R 1985 *J. Biol. Chem.* **260** 4975–83
[13] Mueller J E, Du S M and Seeman N C 1991 *J. Am. Chem. Soc.* **113** 6306–8
[14] Tang J, Du N and Doyle P S 2011 *Proc. Natl. Acad. Sci. U. S. A.* **108** 16153–8
[15] Arai Y, Yasuda R, Akashi K, Harada Y, Miyata H, Kinosita K and Itoh H 1999 *Nature* **399** 446–8
[16] Mansfield M L 1994 *Nat. Struct. Biol.* **1** 213–4
[17] Takusagawa F and Kamitori S 1996 *J. Am. Chem. Soc.* **118** 8945–6
[18] Taylor W R 2000 *Nature* **406** 916–9
[19] Virnau P, Mirny L A and Kardar M 2006 *PLoS Comput. Biol.* **2** 1074–9

[20] Bölinger D, Sułkowska J I, Hsu H P, Mirny L A, Kardar M, Onuchic J N and Virnau P 2010 *PLoS Comput. Biol.* **6**

[21] King N P, Jacobitz A W, Sawaya M R, Goldschmidt L and Yeates T O 2010 *Proc. Natl. Acad. Sci. U. S. A.* **107** 20732–7

[22] Virnau P, Mallam A and Jackson S 2011 *J. Phys.: Condens. Matter* **23** 033101

[23] Kolesov G, Virnau P, Kardar M and Mirny L A 2007 *Nucleic Acids Res.* **35** W425–8

[24] Jamroz M, Niemyska W, Rawdon E J, Stasiak A, Millett K C, Sułkowski P and Sulkowska J I 2015 *Nucleic Acids Res.* **43** D306–14

[25] Forgan R S, Sauvage J P and Stoddart J F 2011 *Chem. Rev.* **111** 5434–64

[26] Dietrich-Buchecker C O and Sauvage J P 1989 *Angew. Chem. Int. Ed. Engl.* **28** 189–92

[27] Vologodskii A V, Lukashin A V, Frank-Kamenetskii M D and Anshelevich V V 1974 *Zh. Eksp. Teor. Fiz.* **66** 2153–2163

[28] Frank-Kamenetskii M D, Lukashin A V and Vologodskii A V 1975 *Nature* **258** 398–402

[29] Michels J P J and Wiegel F W 1982 *Phys. Lett. A* **90** 381–4

[30] Koniaris K and Muthukumar M 1991 *Phys. Rev. Lett.* **66** 2211–4

[31] Janse van Rensburg E J, Sumners D A W, Wasserman E and Whittington S G 1992 *J. Phys. A* **25** 6557–66

[32] Rybenkov V, Cozzarelli N and Vologodskii A 1993 *Proc. Natl. Acad. Sci. USA* **90** 5307–11

[33] Mansfield M L 1994 *Macromolecules* **27** 5924–6

[34] Grosberg A Y 2000 *Phys. Rev. Lett.* **85** 3858–61

[35] Katritch V, Olson W K, Vologodskii A, Dubochet J and Stasiak A 2000 *Phys. Rev. E* **61** 5545–9

[36] Moore N T, Lua R C and Grosberg A Y 2004 *Proc. Natl. Acad. Sci. U. S. A.* **101** 13431–5

[37] Marcone B, Orlandini E, Stella A L and Zonta F 2005 *J. Phys. A* **38** L15–L21

[38] Orlandini E, Tesi M C and Whittington S G 2005 *J. Phys. A* **38** L795–L800

[39] Virnau P, Kantor Y and Kardar M 2005 *J. Am. Chem. Soc.* **127** 15102–6

[40] Dobay A, Dubochet J, Millett K, Sottas P E and Stasiak A 2003 *Proc. Natl. Acad. Sci. U. S. A.* **100** 5611–5

[41] Micheletti C, Marenduzzo D, Orlandini E and Sumners D W 2008 *Biophys. J.* **95** 3591–9

[42] Reith D, Cifra P, Stasiak A and Virnau P 2012 *Nucleic Acids Res.* **40** 5129–37

[43] Trefz B, Siebert J and Virnau P 2014 *Proc. Natl. Acad. Sci. U. S. A.* **111** 7948–51

[44] Marenz M and Janke W 2016 *Phys. Rev. Lett.* **116**(12) 128301

[45] Wüst T, Reith D and Virnau P 2015 *Phys. Rev. Lett.* **114**

[46] Clisby N 2010 *Phys. Rev. Lett.* **104** 055702

[47] Koniaris K and Muthukumar M 1991 *J. Chem. Phys.* **95** 2873–81

[48] Wittmer J P, Beckrich P, Johner A, Semenov A N, Obukhov S P, Meyer H and Baschnagel J 2007 *Europhys. Lett.* **77** 56003

[49] Fisher M E and Sykes M F 1959 *Phys. Rev.* **114** 45–58

[50] Noonan J 1998 *J. Stat. Phys.* **91** 871–888

[51] Madras N and Slade G 1993 *The Self-Avoiding Walk* (Birkhäuser, Boston)

[52] Fitzner R and van der Hofstad R 2013 *J. Stat. Phys.*

[53] Lal M 1969 *Mol. Phys.* **17** 57–64

[54] Madras N and Sokal A D 1988 *J. Stat. Phys.* **50** 109–186

[55] Kennedy T 2002 *J. Stat. Phys.* **106** 407–429

[56] Clisby N 2010 *J. Stat. Phys.* **140** 349–392

[57] Alexander J W 1928 *T. Am. Math. Soc.* **30** 275–306

Simulation and visualization of crystal shapes and interfaces

Joan Adler, Lilach Saltoun[1] and Emil Polturak
Physics Department, Technion-Israel Institute of Technology, Haifa, Israel.
E-mail: phr7ja@tx.technion.ac.il

Abstract. Nature often arranges atoms in the shape of perfect crystals, but sometimes she creates defects and multiple domains. The optimal crystal shape at zero kelvin can be found via the Wulff construction, which can be only be carried out analytically for those trivial cases where next nearest neighbour interactions can be neglected. For our system of interest - videlicit the HCP non-Bravais case, numerical simulation is needed. This system is of relevance because we are modeling helium crystals. We have modeled two adjacent crystallites with different orientations in contact creating twist or tilt grain boundaries, and calculated the surface energy of the interface. Experience gained from several aspects of this project have a wider applications, including the condensed matter simulation application to sample construction for multi-domain crystals, and a visualization one for representation in the presence of grain boundaries. The optimization of sample shapes into their groundstates is also related to wavefront optimization in multimirror telescopes.

1. Introduction

There is a shift in materials simulations from studying ideal single crystals to crystals made of several grains with distinct orientations, and boundaries between them. This change in paradigm has resulted in interest in automated approaches to sample preparation. One general approach is that of the EU FP7 SimPhoNy project, where different codes are connected by wrappers and a formalism enabling easy interoperability, [1]. Another is the preparation of specific codes such as that of P. Lazic's code (CELLMATCH) to combine two unit cells into a supercell for a graphene system, and its substrate for Density Functional Theory (DFT) calculations with VASP, [2]. Of course such sample preparation can be carried out within Computer Aided Design (CAD) codes such as nCAD by Sgenia (part of SimPhoNy), but unlike CELLMATCH and the code we will introduce below, these do not include the simulation process.

In order to understand experiments on helium crystals, we needed to prepare crystals with several grains in order to study grain boundaries in helium HCP crystals. We are interested in boundaries between grains whose orientations are tilted or twisted. Thus we need to prepare adjacent crystal grains with different types of boundaries between them. The present study calculations HCP crystal shapes, cuts the crystals and then recombines them with tilt ed or twisted boundaries. A full account is provided on the project website [3] and additional material is in a supplementary pdf [4].

[1] Present address: Hebrew University, Jerusalem, Israel.

2. Background

2.1. Experiments

Images of the evolution of a helium crystal on heating from the laboratory of S.G. Lipson at the Technion were collated by JA, published in [5] and presented in [4]. The crystal edges become round and facets become rough as the temperature increases. The change from flat facets with sharp edges to rough and rounded is known as the roughening transition. All crystals exhibit such behaviour, but due to equilibration times it is harder to observe in metals.

2.2. Wulff construction at T=0

Crystals at T=0 have an equilibrium shape with sharp edges and flat facets. This shape can be calculated by the Wulff construction which gives the minimal shape formed by the planes that have the minimal distance from the origin at each angle, [6]. The procedure is as follows: Consider a circular surface element, Δs, at angle θ, ϕ with a normal $\hat{n} = (\sin(\theta), \cos(\phi), \sin(\theta), \cos(\theta))$. A missing bond $\hat{\epsilon}$ would be contained in a unit circle if, the bond energy, $b_i(\theta, \phi) = 1$. We note that $b_i(\theta, \phi) = 1$ if $\hat{n} \cdot \hat{\epsilon} > 0$ and 0 if $\hat{n} \cdot \hat{\epsilon} \leq 0$. The sum of all bonds in Δs for the same b_i is: $N_b(\theta, \phi) = \sum_i b_i$. The surface tension for this shape is:

$$\gamma(\theta, \phi) = \frac{\sum_b N_b(\theta, \phi)\sigma_b}{\Delta s},$$

where σ_b is the bond energy. The Wulff construction is analytic for simple Bravais nearest neighbour cases, and can be done numerically otherwise, with details given in [5].

Research into roughening transitions - Theoretical, Experimental and computational has been ongoing at the Technion for 40 years. Several faculty and many graduate students have been involved - Avron, Landau, Lipson, Adler, Balfour, Carmi, Baum, Geminterm, Hashibon, Polturak and most recently Saltoun.

2.3. The HCP lattice

The Hexagonal Close Packed (HCP) lattice is non-Bravais and like the other common non-Bravais lattice, that of diamond, is common to many interesting elements and compounds. HCP structures include one allotrope of solid helium, and the metal magnesium. Numerical studies on HCP are complicated by the two sublattice nature, and it requires 4 Miller indices. A discussion of this structure was given by two of us in a study of melting in magnesium [7]. The rich structure of HCP solid Helium makes the additional effort worthwhile again here. Fig. 1 shows the lattice structure at left.

3. Technion simulations of roughening transitions and their visualization

Crystal interfaces and their roughening can be studied numerically with SOS (Solid-On-Solid) models, or as interfaces in Ising models, either by direct simulation or for simple cases by series expansions. The SOS models have also been applied to telescope phasing. For the simulations, visualization is very important especially to check the complicated boundary conditions. Early examples are presented in [4].

The surface tension of the Next Nearest Neighbour (NNN) HCP is shown in the central frame of Fig. 1 (Mathematica Image by G. Baum, published in [5]) where the indentations indicate facets. After the initial observation of the roughening transition, interest turned to the order of roughening temperatures of the different facets. The HCP crystal shape is shown in Fig. 2 at upper left, with the 3 facets, known as a, c, and s, in the NN version and at lower left in the NNN version which has additional facets, cs and a2. In the central panel of Fig. 2 we show the Wulff surface with outlines of the crystal facets inside. The additional facets were researched with a simulation of surfaces of Ising models in the MSc thesis of A. Hashibon. We concluded that: A large NNN term was needed in order to achieve experimental/simulation agreement

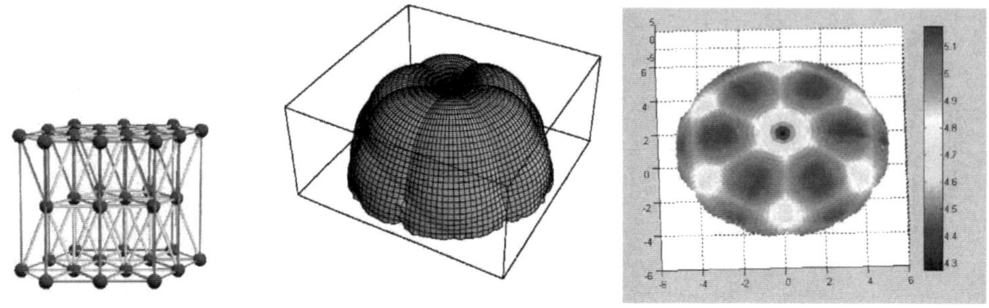

Figure 1. At left, the HCP lattice by AViz, center, the Wulff surface by G. Baum and right, Wulff surface from this project.

between the ratios of the roughening transition temperatures of the different facets. This could be realistic, or the experimental measurements could not have been in true equilibrium. To the best of our knowledge this was not confirmed, because experimental focus turned to different directions, such as the present calculations.

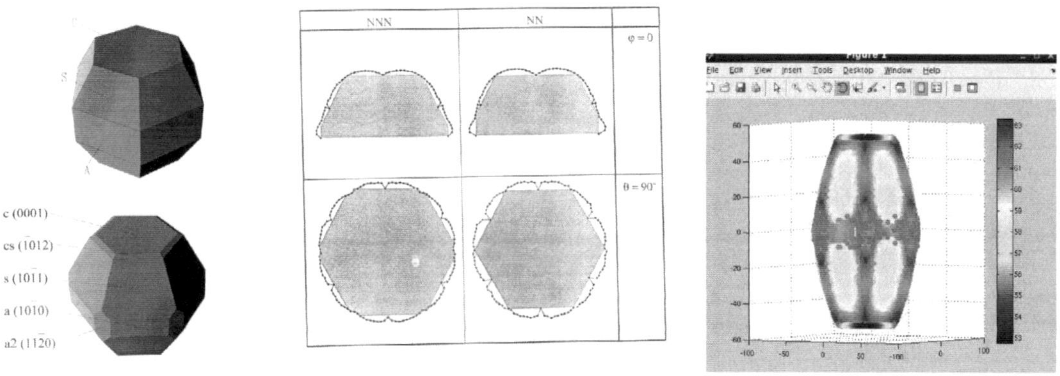

Figure 2. Crystal shape with only NN interactions at upper left and with NNN interactions at lower left and a 2D cut of the crystal creation from the wulf plot at right.

4. Cutting and recombining crystals

The code used for the calculation of the surface free energy was originally published in the appendix of G. Baum's MSc. thesis, Technion, 1994, and restored from backups to be recycled for the present project, and took about 5 mins to recompile in LINUX gfortran after 20 years. Once we could again create crystals with the Wulff shape we began to cut and recombine in order to study the boundaries. The Wulff surface and crystal crystal shape from the present calculation are shown in the rightmost panels of Figs. 1 and 2 providing validation.

4.1. Steps in the process

Our process has several steps, whose details are presented in [4]. After the execution of an initial fortran code that calculates the surface tension, the next stages of the calculations were carried out with interactive Graphical User Interfaces (GUIs) in MATLAB. These stages included using the surface tension to calculate the Wulff shape and create an HCP crystal. This crystal was then cut, into crystal1 and crystal2, and the resulting crystals rotated and then recombined with

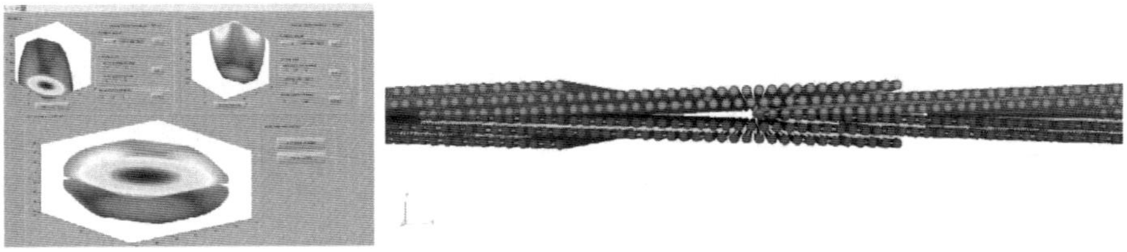

Figure 3. Interface of the cutting_rot_and_conecting_crystals.m routine at left and a closeup of the tilted interface at right.

different angles of twist or tilt. The surface energy was calculated for a selection of angles and boundary crystals. The final visualization was done with AViz [8]. The stages can be itemized as:

(i) Select sample size, a typical choice would be a length, L of 50.
(ii) Calculate the surface tension of an HCP crystal in the next nearest neighbour approximation.
(iii) Calculate the Wulff shape.
(iv) Create an HCP crystal in the Wulff shape.
(v) Cut, rotate and connect crystals.
(vi) Calculate surface energy for twisted boundaries, with a selection of boundary conditions.
(vii) Calculate surface energy for tilted boundaries, with a selection of boundary conditions.
(viii) Visualize with AViz.

4.2. The MATLAB and AViz codes

There are five distinct GUI-enabled MATLAB codes for these processes, each with its own interface. They are described in detail on the website [3], and summarised in [4]. We show one example, the central cutting_rot_and_conecting_crystals.m GUI routine in Fig. 3. Datafiles of crystal boundary coordinates were found and then fed into surface energy calculation routines. We generated .xyz files as another interactive option and a tilted interface is shown on the right of Fig. 3 drawn with AViz [8].

4.3. Estimation of surface energy

The main application of the new code is the calculation of the boundary surface energy. Three such graphs as functions of twist and tilt direction are presented in Fig. 4, with the leftmost panel's boundary conditions (BC) being infinite, and the center's finite. These relate to whether or not the outer contour surface is ignored when calculating bond projections to approximate cases of missing bonds.

The parameters 0.5 v. 085 relate to a threshold for "close enough" to decide if a bond is present in the infinite boundary case. In the tilted case only 0.5 was used. In both twisted cases surface energy close to zero when rotation is zero. In the infinite case surface energy close to constant because no. of bonds do not change a lot while rotating. In the tilted case a much stronger angular dependence is seen.

4.4. Interface visualization with AViz

A very "transparent" way to categorise the structure of the interface is to use AViz for several layers on both sides, and to draw in bonds between atoms in adjoining crystals. The uncut is

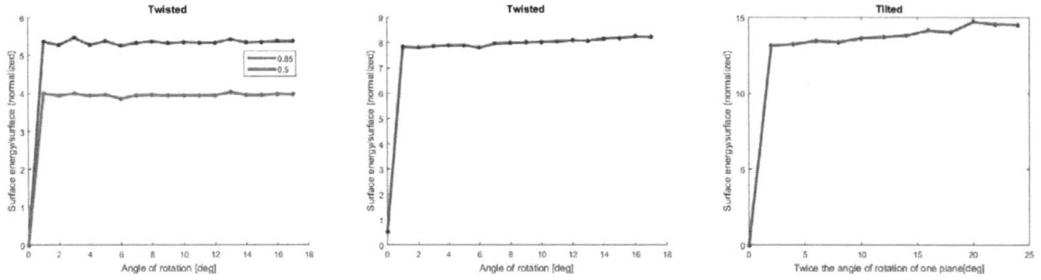

Figure 4. Surface energy results as functions of twist and tilt angles.

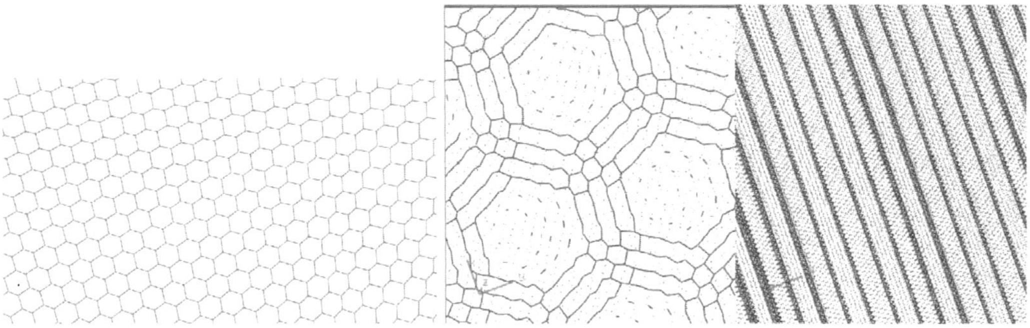

Figure 5. Interfaces, L-R uncut, twisted and tilted.

translationally invariant, the twisted has complex structure and the tilted has simple stripes. These are shown, left to right, in Fig. 5.

Acknowledgments

This study was carried out as a final undergraduate project. The earlier calculations and related experimental studies were supported by a series of ISF grants to JA and EP. The AViz visualizations are supported by the SimPhoNy project which is funded by FP7 under NMP-2013-1.4-1 call with grant agreement no: 604005.

References

[1] `http://www.simphony-project.eu`

[2] Lazik P 2015 Combining two unit cells into a common supercell wuth minimal strain *Computer Physics Communications* **197** 324

[3] `http://phony1.technion.ac.il/~lilach`

[4] `http://phony1.technion.ac.il/~phr76ja/ath16/supplement.pdf`

[5] Hashibon A, Adler J, Baum G, and Lipson S G 1998 Roughening transitions and surface tension in an hcp lattice with higher neighbor interactions *Phys. Rev. B* **58** 4120

[6] Herring C 1951 Some Theorems on the Free Energies of Crystal Surfaces *Phys. Rev.* **82** 87

[7] Bavli P, Polturak E and Adler J 2001 Molecular dynamics study of melting in the hcp metal Mg *Phys. Rev. B* **84** 235442

[8] Peled D, Silverman A and Adler J 2013 3D visualization of atomistic simulations on every desktop *IOP Conference Series* **454** 012076

Journal of Physics: Conference Series **750** (2016) 012012 doi:10.1088/1742-6596/750/1/012012

Phase transitions of 2D triangular antiferromagnetic Ising model in a uniform magnetic field near h=0 and T=0

Wooseop Kwak[1], Sanghyun Kim[2], and Chi-Ok Hwang[3*]

[1]Department of Physics, Chosun University, Gwangju, 61452, Republic of Korea
[2]Concentration of Physics, GIST College, Gwangju Institute of Science and Technology, Gwangju, 61005, Republic of Korea
[3]Division of Liberal Arts and Sciences, GIST College, Gwangju Institute of Science and Technology, Gwangju, 61005, Republic of Korea

*chwang@gist.ac.kr

Abstract. Using a microcanonical magnetization–energy (ME) diagram, this paper describes all the possible phase transitions of a 2D triangular antiferromagnetic (TAFM) Ising model in a uniform external magnetic field. In particular, we investigate the detailed phase boundary shape of the TAFM near h=0 and T=0 using staggered susceptibility.

1. Introduction

The 2D isotropic triangular antiferromagnetic (TAFM) model with no magnetic field has been solved fully. It has highly degenerate ground states, and there is no disorder–order phase transition [1]. However, an external magnetic field induces a long-range order with one-third magnetization because of the frustration. In the long-range order, there are two sublattices aligned to the field (called the α sublattices) and one sublattice (called the β sublattice) anti-aligned to the field [2] (see Figure 1). Here, we consider the triangular-lattice Ising antiferromagnet in a uniform external magnetic field, with the Hamiltonian $H = -J\sum_{<i,j>}\sigma_i\sigma_j - h\sum_{i=1}^{N}\sigma_i$ ($J < 0$). $< i, j >$ denotes distinct pairs of nearest-neighbor sites.

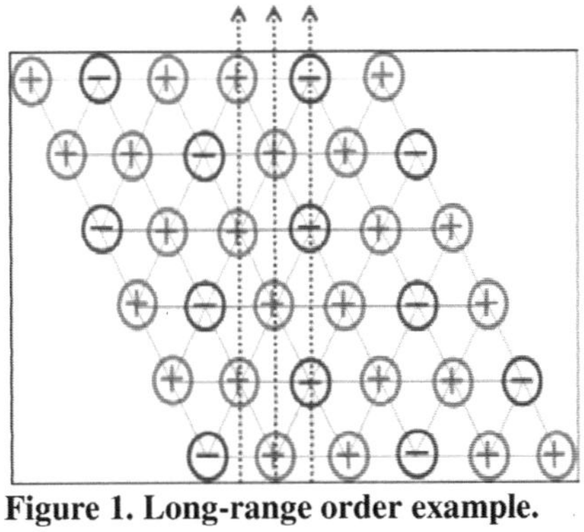

Figure 1. Long-range order example.

In [3-6], it was found that there is a field-induced Kosterlitz–Thouless (KT) transition to a long-range-ordered state at a field H_c =0.266+0.010 at the limit of h → 0 and T−> 0, keeping H = h/T. a constant. Even though this 2D frustrated system is a classical problem, and many advanced frustrated systems have been well investigated [7], it still seems that we do not understand the classical problem completely yet. Besides the KT transition, there is another phase transition belonging to the three-state Potts universality class [8]. However, around the point h = 0 and T = 0, we do not know where the three-state Potts transition starts, because the crossover effect degrades the convergence. In addition, we do not know why we have two different transition types on the TAFM, because from the microcanonical ME diagram, we can expect only one type. In addition, if we use another order parameter, we do not know what will happen (see Figure 4).

Besides the usual total magnetization order parameter $\sum_i \sigma_i$, we can use another order parameter, so-called staggered magnetization $\sum_{i\in\beta} \sigma_i - \sum_{j\in\alpha} \sigma_j$, so that the order parameter ranges from -1 to 1. If we use this order parameter, in any given plaquette, two sublattices will have the same average value of spin, and the order parameter will be a 2D one made up of appropriate combinations of the sublattice magnetizations. Hence, the ground state in a magnetic field is degenerate. A long time ago in a previous research study, a phase diagram was obtained using staggered magnetization [2].

In our previous research [9], we obtained the density of states (DOS) g(M,E) as a function of the energy (E) and magnetization (M) of the TAFM using the exact enumeration method for small systems and the Wang–Landau method for larger systems (see Figure 2). From the top view of the DOS, we could obtain the magnetization–energy (ME) diagram (see Figure 3).

In this paper, on the microcanonical ME diagram, we describe all the possible phase transitions of the 2D TAFM Ising model in a uniform magnetic field and investigate the detailed phase boundary shape of the TAFM near h=0 and T=0.

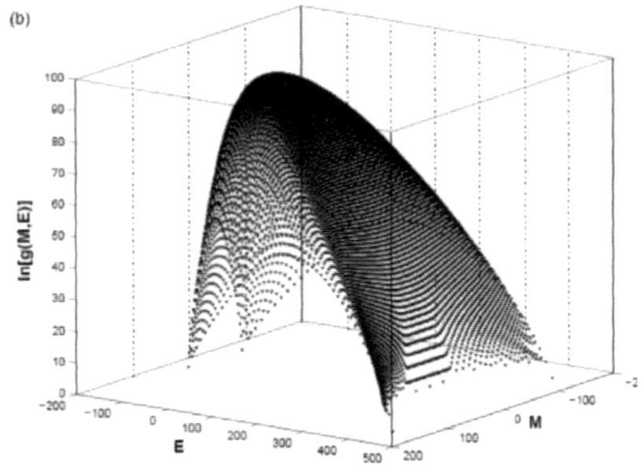

Figure 2. Example of density of states (DOS) with size L = 10 [5].

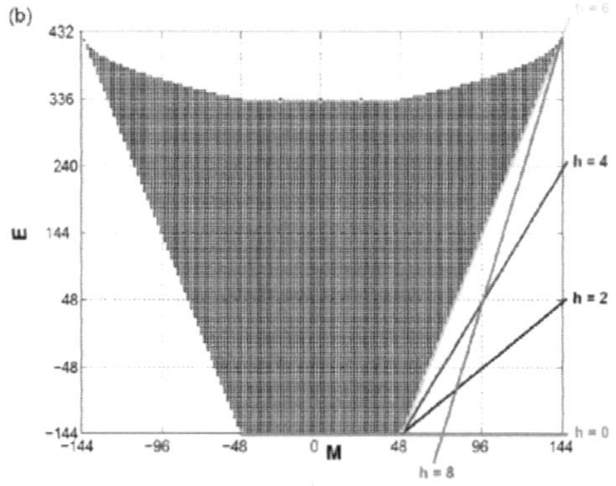

Figure 3. Magnetization–Energy (ME) diagram of TAFM with size 12x12 [5]. Solid colored lines are shown for the ground states with external magnetic fields.

2. Phase transitions of ground states

To understand the phase transitions of ground states, it is very useful to use the ME diagram obtained by Wang–Landau sampling. From the ME diagram, we can understand all the ground states in external magnetic fields. From the Hamiltonian $E_t = E - hM$, we can notice that the external magnetic field is the slope on the ME diagram. On the diagram, we can clearly see that the external magnetic field h = 6 is the critical magnetic field, where the ground state can be mapped onto the Baxter's hard-hexagon lattice gas [10], and there are four critical ground–state points: $E_{min} = L^2$ and M = 0 when h = 0, $E_{min} = L^2$ and M = $L^2/3$ when 0 < h < 6 (long-range order), E = L^2 and M = $2L^2/3$ when h = 6, and E = $3L^2$ and M = L^2 when h > 6 (here, L is the lattice size). According to the given field change, such as from h = 0 to any magnetic field in 0 < h < 6, we can see a phase transition from one ground state to another.

3. Phase boundary shape of TAFM near h=0 and T=0

The field-induced KT transition at the limit of h → 0 and T−> 0 keeping H = h/T. a constant can be interpreted as the tangent slope of the phase boundary at h=0 and T=0. The tangent slope at the point is not zero, and it means that with a nonzero small external magnetic field at the ground state, the system does not go to the long-range order immediately but the magnetic field strength is marginal (see Figure 5). In our previous research [11] we confirmed the critical value H_c using normal magnetization susceptibility with Wang–Landau sampling (see Figure 6).

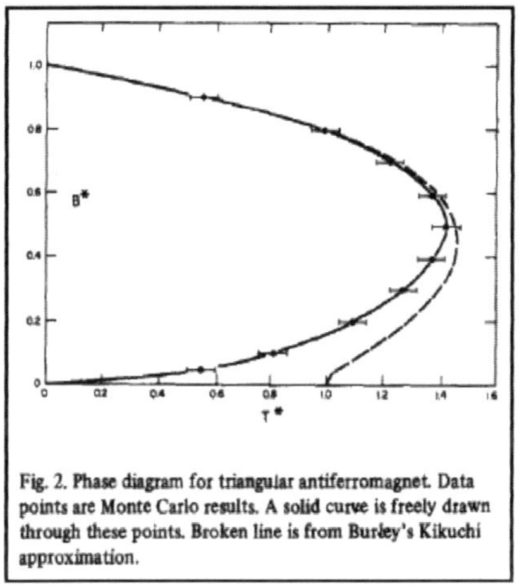

Figure 4. Phase diagram obtained by Metcalf [2].

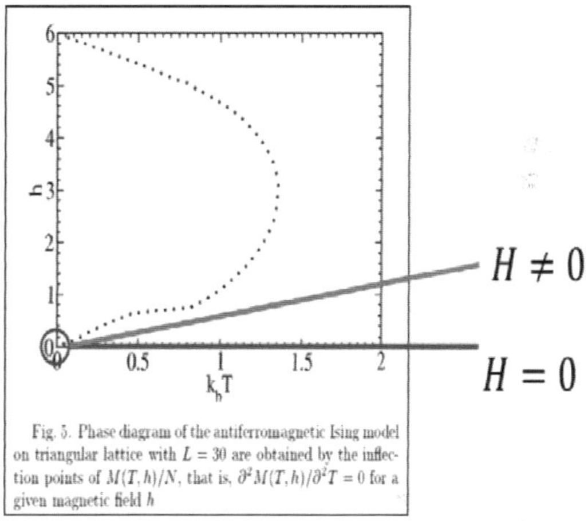

Figure 5. Phase diagram from [5].

Figures 4 and 5 are different, because they use different order parameters. The two magnetic lines in Figure 5 show that with a nonzero small external magnetic field at the ground state, the system does not go to the long-range order immediately [5].

For staggered susceptibility, we can use the variance of $\sum_{i\in\beta}\sigma_i - \sum_{j\in\alpha}\sigma_j$as an ordering parameter. In this case, we need the more expanded DOS, g(M_1, M_2, E), where $M_1 = \sum_{i\in\beta}\sigma_i$ and $M_2 = \sum_{j\in\alpha}\sigma_j$. We had difficulty using Wang–Landau sampling because of the high dimension of the DOS and thus used Metropolis sampling to obtain Figures 7 and 8. We started the TAFM from the long-range order state and slowly increased the temperature. After reaching the equilibrium, we collected the data. In Figure 7, we have two peaks corresponding to the KT phase transition (right peak) and another boundary, we suspect, related to Burley's Kikuchi approximation (left peak) [12]. It seems that the two peaks diverge with respect to the linear system size. In future research, to take full advantage of Metropolis sampling, we will perform simulations with much bigger sizes up to the linear system sizes, such as 513 and 1026.

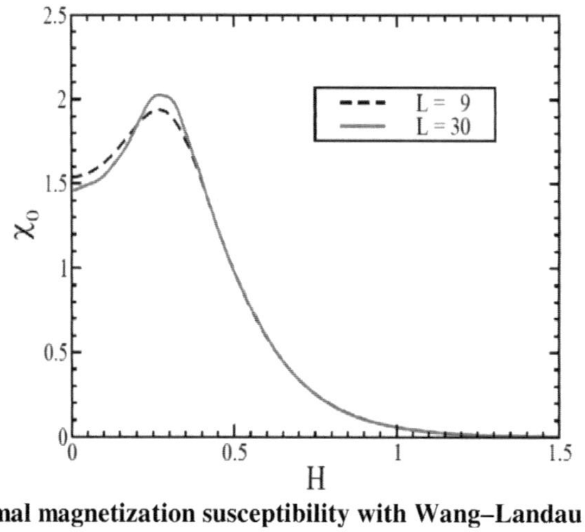

Figure 6. Normal magnetization susceptibility with Wang–Landau sampling [12].

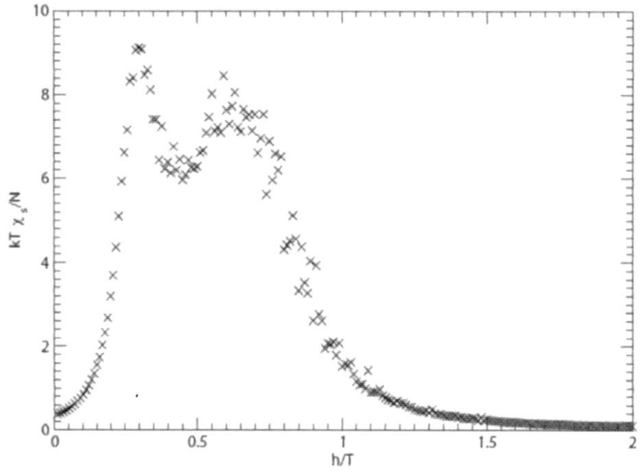

Figure 7. Staggered susceptibility with L = 30 and h = 0.2 using Metropolis algorithm.

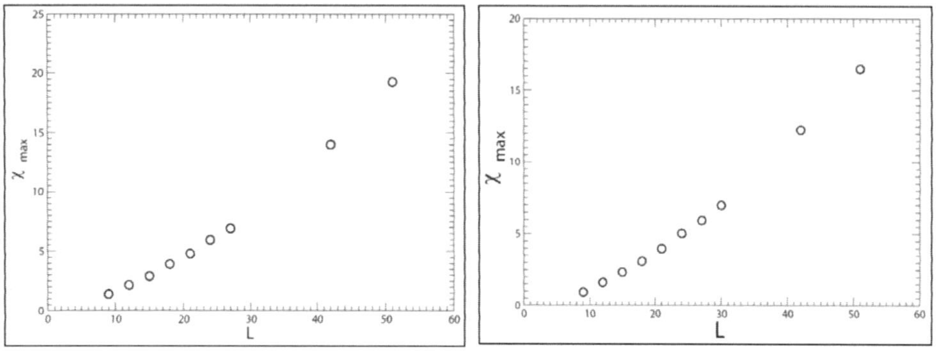

Figure 8. Staggered susceptibility maxima vs. lattice sizes.

4. Conclusion and further research

In this research, we used the ME diagram to understand all the possible phase transitions at ground states, and we investigated the detailed phase boundary shape of the TAFM near h=0 and T=0 using staggered susceptibility and confirmed the KT transition at the limit of $h \rightarrow 0$ and $T -> 0$. However, in our preliminary research results, we could also see another phase boundary line.

In our future research, we will investigate the phase boundary with bigger linear system sizes. In addition, we will investigate the phase transitions via microcanonical entropy inflection points. We hope to determine where the two phase transitions cross over on the phase boundary.

References

[1] Wannier G. H. 1950 *Phys. Rev.* 79 357
[2] Metcalf B. D. 1973 *Phys. Lett.* 45A(1) 1
[3] Blöte H. W. J. and Nightingale M. P. 1993 *Phys. Rev. B* 47, 15046
[4] Qian X., Wegewijs M. and Blöte H. W. J. 2004 *Phys. Rev. E* 69 036127
[5] Qian X. and Blöte H. W. J. 2004 *Phys. Rev. E* 70 036112
[6] Queiroz S. L. A., Paiva T., Martins J. S., and Santos R. R. 1994 *Phys. Rev. E* 59(3) 2772
[7] Kawamura H. 1998 *J. Phys.: Condens. Matter* 10 4707
[8] Noh J. D. and Kim D. 1992 *Int. J. Mod. Phys. B* 6(17) 2913
[9] Hwang C.-O., Kim S.-Y., Kang D., and Kim J. M. 2009 *J. Stat. Mech.* 5 L05001
[10] Baxter R. J. 1980 *J. Phys.* A13 L61
[11] Hwang C.-O. and Kim S.-Y. 2014 *Monte Carlo Methods and Applications* 20(3) 217
[12] Burley D. M. 1965 *Proc. Phys. Soc.* 85 1163

Acknowledgments

This research was supported by "Research support projects for distinguished professors" through a grant provided by GIST in 2016 (Grant No. GK03872 for C.-O. Hwang) and by the Basic Science Research Program through the National Research Foundation of Korea (NRF) funded by the Ministry of Education, Science, and Technology (NRF2013R1A1A2065043 for W. Kwak). In addition, this research was performed as a collaborative research project of project No (Supercomputing infrastructure service and application for C.-O. Hwang) and supported by the Korea Institute of Science and Technology Information (KISTI).

Irreversible simulated tempering algorithm with skew detailed balance conditions: a learning method of weight factors in simulated tempering

Yuji Sakai[1] **and Koji Hukushima**[1,2]

[1] Department of Basic Science, Graduate School of Arts and Sciences, The University of Tokyo, 3-8-1 Komaba, Meguro-ku, Tokyo, 153-8902, Japan

[2] Center for Materials Research by Information Integration, National Institute for Materials Science, 1-2-1 Sengen, Tsukuba, Ibaraki 305-0047, Japan

E-mail: yuji0920@huku.c.u-tokyo.ac.jp

Abstract. Recent numerical studies concerning simulated tempering algorithm without the detailed balance condition are reviewed and an irreversible simulated tempering algorithm based on the skew detailed balance condition is described. A method to estimate weight factors in simulated tempering by sequentially implementing the irreversible simulated tempering algorithm is studied in comparison with the conventional simulated tempering algorithm satisfying the detailed balance condition. It is found that the total amount of Monte Carlo steps for estimating the weight factors is successfully reduced by applying the proposed method to an two-dimensional ferromagnetic Ising model.

1. Introduction

Since the invention of the Markov-chain Monte Carlo (MCMC) method [1], some improvement and development have been made for an efficient sampling, such as extended-ensemble methods [2, 3, 4, 5] and cluster algorithms [6, 7], and they have been developed within the framework of the detailed balance condition (DBC). Recently, the MCMC methods without the DBC have been discussed to improve their performance and several MCMC algorithms violating the DBC have been reported [8, 9, 10, 11, 12]. In addition, the efficiency of MCMC algorithms without the DBC is partly confirmed theoretically, in terms of the relaxation rate [13] and the asymptotic variance [14].

In the present paper, we make a brief review on a simulated tempering algorithm violating the DBC called the irreversible simulated tempering, particularly by imposing a skew detailed balance condition (SDBC), instead of the DBC. The simulated tempering algorithm has a specific weight factor to be fixed in advance, which determines the efficiency of the algorithm. We also discuss the way to estimate weight factors in the simulate tempering by the sequential update with the simulated tempering algorithm with the SDBC.

2. Irreversible simulated tempering algorithm

In this section, the outline of the irreversible simulated tempering (IST) algorithm based on the lifting technique with the skew detailed balance condition [15] is briefly explained.

Let $\boldsymbol{X}$ be a configuration and $E(\boldsymbol{X})$ denotes an energy function of a system to be considered. In statistical physics, the target distribution of the system is often given by the Gibbs-Boltzmann distribution with an inverse temperature β. In the simulated tempering, the inverse temperature is treated as a random variable as well as the configuration $\boldsymbol{X}$. More specifically, β takes R different values $\{\beta_r\}_{r=1}^{R}$ that should be determined before simulation. In addition, a lifting variable $\varepsilon \in \{+,-\}$ is introduced to the system in the irreversible simulated tempering. Thus, a state in the irreversible simulated tempering is specified by $(\boldsymbol{X}, \beta_r, \varepsilon)$. Accordingly, the target distribution is given as

$$P_{\rm IST}(\boldsymbol{X}, \beta_r, \varepsilon) \propto \exp[-\beta_r E(\boldsymbol{X}) + g_r], \tag{1}$$

where g_r denotes a weight factor depending only on the inverse temperature.

An explicit update scheme of the irreversible simulated tempering algorithm consists of two steps. one is the update scheme of an original configuration $\boldsymbol{X}$ for a fixed inverse temperature and lifting variable with a conventional MCMC algorithm such as the Metropolis-Hastings algorithm and cluster algorithms. The other is the update scheme of the inverse temperature and the lifting variable for fixed $\boldsymbol{X}$, described as follows:

(a) Let the current state be $(\boldsymbol{X}, \beta_r, \varepsilon)$ and the candidate of the next inverse temperature β_l is determined with the probability $q_{r,l}^{(\varepsilon)}$ given as follows:

$$q_{1,2}^{(\varepsilon)} = q_{R,R-1}^{(\varepsilon)} = 1, \tag{2}$$

$$q_{r,r\pm1}^{(\varepsilon)} = \frac{1 \pm \delta\varepsilon}{2}, \tag{3}$$

for $1 < r < R$, and $q_{r,l}^{(\varepsilon)} = 0$ otherwise.

(b) Accept the next state $(\boldsymbol{X}, \beta_l, \varepsilon)$ with the probability $W_{r,l}^{(\varepsilon)}$ given by

$$W_{r,l}^{(\varepsilon)} = \min\left[1, \frac{q_{l,r}^{(-\varepsilon)}}{q_{r,l}^{(\varepsilon)}} \frac{P_{\rm IST}(\boldsymbol{X}, \beta_l, -\varepsilon)}{P_{\rm IST}(\boldsymbol{X}, \beta_r, \varepsilon)}\right]. \tag{4}$$

(c) If the trial (b) is rejected, flip the lifting variable ε with the probability $\Lambda_r^{(\varepsilon)}$ given by

$$\Lambda_r^{(\varepsilon)} = \max\left[0, \varepsilon \sum_{\varepsilon'=\pm} \sum_{l \neq r} \varepsilon' q_{r,l}^{(-\varepsilon')} W_{r,l}^{(-\varepsilon')}\right] \Big/ \left(1 - \sum_{l \neq r} q_{r,l}^{(\varepsilon)} W_{r,l}^{(\varepsilon)}\right) \tag{5}$$

and set $(\boldsymbol{X}, \beta_r, -\varepsilon)$ as the next state.

(d) If the trial (c) is also rejected, set the current state as the next state.

Note that the acceptance probability satisfies the SDBC with respect to the target distribution $P_{\rm IST}(\boldsymbol{X}, \beta_r, \varepsilon)$ and the global balance condition is fulfilled in the above procedure. The parameter δ in the proposal probability $q_{r,l}^{(\varepsilon)}$ controls the violation of the DBC. When δ is set to zero, the DBC is restored.

In Ref. [15], the irreversible simulated tempering algorithm has been applied to the two-dimensional ferromagnetic Ising model as a benchmark. It is numerically shown that the relaxation dynamics of the inverse temperature qualitatively changes from diffusive to ballistic by violating the DBC and consequently the autocorrelation time of the magnetization is reduced several times compared to the conventional simulated tempering for the case with an ideal choice of the weight factors. Thus, it is confirmed that the violation of the DBC can improve the efficiency of simulated tempering algorithm.

Journal of Physics: Conference Series **750** (2016) 012013 doi:10.1088/1742-6596/750/1/012013

3. On-the-fly estimation of weight factors

The choice of the weight factors essentially affects the efficiency of the simulated tempering. When $g_r = -\ln Z(\beta_r)$ with $Z(\beta)$ being the partition function at β, the marginal probability of inverse temperature of the target distribution is uniform and all the inverse temperature is expected to be sampled uniformly. However, it is difficult to know the exact value of the partition function a priori in general. Then, in practice the weight factors are prepared approximately by some algorithms [16, 17, 18, 19, 20]. In this section, we study an on-the-fly estimation method of the optimal weight factors with the irreversible simulated tempering algorithm. The basic idea of the on-the-fly scheme has already proposed in Ref. [19].

Let $(\boldsymbol{X}^{(n)}, \beta^{(n)}, \varepsilon^{(n)})$ be the state of the irreversible simulated tempering after n MCS and $N_r^{(n)}$ denotes the histogram of the inverse temperature β_r defined as

$$N_r^{(n)} \equiv \sum_{k=1}^{n} \delta(\beta^{(k)}, \beta_r), \tag{6}$$

where $\delta(\cdot,\cdot)$ is the Kronecker delta. Then, the estimator of expectation of energy at $\beta = \beta_r$ is given as

$$\bar{E}_r^{(n)} \equiv \frac{1}{N_r^{(n)}} \sum_{k=1}^{n} E(\boldsymbol{X}^{(k)}) \delta(\beta^{(k)}, \beta_r), \tag{7}$$

and the weight factors are updated iteratively according to the following equation,

$$g_{r+1} = g_r + \frac{\bar{E}_r^{(n)} + \bar{E}_{r+1}^{(n)}}{2} (\beta_{r+1} - \beta_r). \tag{8}$$

The on-the-fly weight determination scheme is given as follows.

(a) Prepare a set of inverse temperatures $\{\beta_r\}_{r=1}^{R}$ such that $\beta_1 > \beta_2 > \cdots > \beta_R$ without loss of generality. Set $g_r^{(0)} = 0$ for $r = 1, 2, ..., R$.

(b) Start the IST simulation from the largest inverse temperature β_1 and a random initial condition $\boldsymbol{X}^{(0)}$ and $\varepsilon^{(0)}$. Calculate $\bar{E}_1^{(n)}$ and update g_2 as $g_2 = (\beta_2 - \beta_1)\bar{E}_1^{(n)}/2$ after every MCS.

(c) If the transition from β_1 to β_2 is realized for the first time, start to calculate $\bar{E}_2^{(n)}$ and update g_2 and g_3 according to Eq. (8).

(d) Continue the IST simulation, calculate $\bar{E}_1^{(n)}$ and $\bar{E}_2^{(n)}$, and update g_2 and g_3 after every MCS. If the transition to β_3 is realized for the first time, start to calculate $\bar{E}_3^{(n)}$ and update g_3 and g_4 according to Eq. (8).

(e) Continue the above procedure until g_R is updated. Note that during the simulation, we update weight factors g_r and g_{r+1} after each MCS when the current inverse temperature is β_r.

We apply the above algorithm to the two-dimensional ferromagnetic Ising model without external magnetic field. In this work, The number of spins is fixed as $N = 32^2$ and the set of inverse temperature is prepared as $\beta_1 = 0.5$, $\beta_R = 0.2$, and the others are aligned with equal interval. A Monte Carlo step (MCS) is defined as the time unit where N spin-flip trials with Metropolis-Hastings algorithm and a update trial of (β, ε) are performed.

First, we study the total number of MCSs required to accomplish the estimation. Figure 1 is showing R-dependence of the total MCSs. While no difference between $\delta = 0$ and $\delta \neq 0$ is

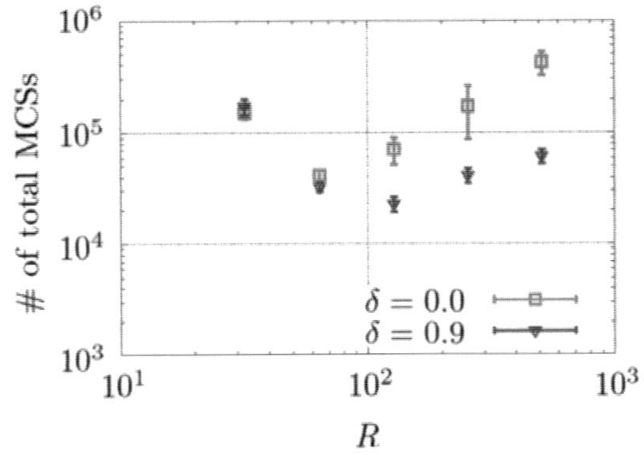

Figure 1. R-dependence of total number of MCSs for the on-the-fly estimation of weight factors. The parameter δ is chosen as $\delta = 0.0$ (red square) and $\delta = 0.9$ (blue inverted triangle), respectively.

observed in small R region in Fig. 1, the total MCSs in the case of $\delta \neq 0$ is about 10 times less than that in the case of $\delta = 0$ when the number of inverse temperature is sufficiently large. It indicates that the computational cost of the estimation of weight factors can be reduced by violating the DBC.

We next focus on the estimation accuracy of weight factors. Figure 2 shows that the weight factors estimated by our algorithm with the parameter $\delta = 0$ and $\delta = 0.9$ are comparable. Thus, we conclude that the violation of the DBC can improve the estimation algorithm proposed in Ref. [19].

4. Summary

In this paper, the irreversible simulated tempering algorithm violating the DBC was briefly reviewed. The update scheme in the simulated tempering algorithm is modified by introducing the lifting technique and the skew detailed balance condition is imposed to the acceptance probability of inverse temperature. It turned out [15] that the violation of the DBC accelerates the relaxation of inverse temperature and the autocorrelation of magnetization in comparison with the conventional simulated tempering. Here we propose an iterative algorithm to estimate the optimal weight factor by implementing the irreversible simulated tempering. By applying the method to the Ising model in two dimensions, it is revealed that the total Monte Carlo steps required in the estimation is successfully reduced without deteriorating the precision of the estimation by violating the DBC. Our results indicate that the violation of the DBC by imposing the SDBC can also improve the efficiency of the estimation of weight factors in simulated tempering. As a future work, our algorithm would be applied to more complicated systems such as the q-state Potts model with large value of q and spin glasses.

acknowledgments

The authors are grateful to S. Todo for useful comments and for letting us notice the method of Ref. [21]. YS is supported by a Grant-in-Aid from the Japan Society for Promotion of Science (JSPS) Fellows (Grant No. 26·7868). KH is supported by Grants-in-Aid for Scientific Research from MEXT, Japan (Grant Nos. 25610102 and 25120010).

References

[1] Metropolis N, Rosenbluth A W, Rosenbluth M N, Teller A H and Teller E 1953 *J. Chem. Phys.* **21** 1087–92
[2] Berg B A and Neuhaus T 1992 *Phys. Rev. Lett.* **68** 9
[3] Hukushima K and Nemoto K 1996 *J. Phys. Soc. Jpn.* **65** 1604
[4] Lyubartsev A P, Martinovski A A, Shevkunov S V and Vorontsov-Velyaminow P N 1992 *J. Chem. Phys.* **96** 1776
[5] Marinari E and Parisi G 1992 *Europhys. Lett.* **19** 451

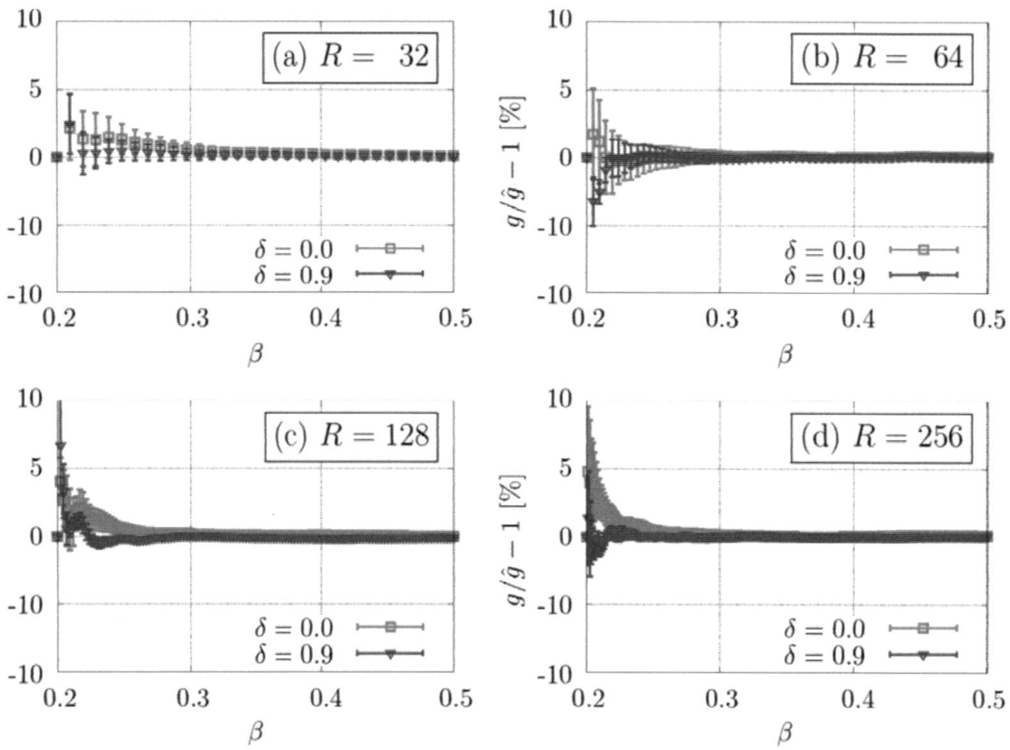

Figure 2. (Color online) Estimation accuracy of weight factors as a function of inverse temperature β. The number of inverse temperature is (a)$R = 32$, (b)$R = 64$, (c)$R = 128$, and (d)$R = 256$. The parameter δ is chosen as $\delta = 0.0$ (red square) and $\delta = 0.9$ (blue inverted triangle). Horizontal axis in each panel represents the ratio between the weight factor g obtained by on-the-fly estimation and the weight factor $\hat{g}$ evaluated by the exact numerical method in Ref. [21].

[6] Swendsen R H and Wang J S 1987 *Phys. Rev. Lett.* **58** 86
[7] Wolff U 1989 *Phys. Rev. Lett.* **62** 361
[8] Suwa H and Todo S 2010 *Phys. Rev. Lett.* **105** 120603
[9] Turitsyn K S, Chertkov M and Vucelja M 2011 *Physica* D **240** 410
[10] Fernandes H C and Weigel M 2011 *Comput. Phys. Commun.* **182** 1856
[11] Schram R D and Barkema G T 2014 *Physica* A **418** 88
[12] Bierkens J 2015 *Stat. Comput.*
[13] Ichiki A and Ohzeki M 2013 *Phys. Rev.* E 88 020101(R)
[14] Sun Y, Gomez F and Schmidhuber J 2010 NIPS2010 proceedings 2235
[15] Sakai Y and Hukushima K 2016 *Preprint* cond-mat.stat-mech/1601.04286
[16] Hansmann U H E and Okamoto Y 1997 *J. Comput. Chem.* **18** 920
[17] Mitsutake A and Okamoto Y 2004 *J. Chem. Phys.* **121** 2491
[18] Park S and Pande V S 2007 *Phys. Rev.* E **76** 016703
[19] Nguyen P H, Okamoto Y and Derreumaux P 2013 *J. Chem. Phys.* **138** 061102
[20] Valentim A, da uz M G E and Fiore C E 2014 *Comput. Phys. Commun.* **185** 2046
[21] Kastening B 2001 *Phys. Rev.* E **64** 066106

Event-chain Monte Carlo algorithm for continuous spin systems and its application

Yoshihiko Nishikawa

Department of Basic Science, The University of Tokyo 3-8-1 Komaba, Meguro, Tokyo 153-8902, Japan

E-mail: `nishikawa@huku.c.u-tokyo.ac.jp`

Koji Hukushima

Department of Basic Science, The University of Tokyo 3-8-1 Komaba, Meguro, Tokyo 153-8902, Japan
Center for Materials Research by Information Integration, National Institute for Materials Science, 1-2-1 Sengen, Tsukuba, Ibaraki 305-0047, Japan

E-mail: `hukusima@phys.c.u-tokyo.ac.jp`

Abstract. The event-chain Monte Carlo (ECMC) algorithm is described for hard-sphere systems and general potential systems including interacting particle system and continuous spin systems. Using the ECMC algorithm, large-scale equilibrium Monte Carlo simulations are performed for a three-dimensional chiral helimagnetic model under a magnetic field. It is found that critical behavior of a phase transition changes with increasing the magnetic field.

1. Introduction

Since the proposal by Metropolis *et al.* [1], Markov-chain Monte Carlo (MCMC) algorithms have been used in a wide range of research fields such as statistics, chemistry, and physics. Particularly, in statistical physics, MCMC algorithms are powerful methods to study finite-dimensional effects of many-body systems. However, some simple algorithms with local updates including the original Metropolis algorithm frequently suffer from difficulties of slow relaxation such as critical slowing down associated with continuous phase transitions, nucleation process in first-order phase transitions, and glassy dynamics in complex systems. These slowing-down problems make Monte Carlo simulations take very long time and the precision of physical quantities be decreased. In order to overcome the slowing down, roughly two types of algorithms have been proposed: the first type is the cluster algorithms [2, 3] and the other one is the extended ensemble algorithm [4–9].

Recently, another type of algorithms has attracted a great interest: that is the Monte Carlo algorithms which break the detailed balance condition, but satisfy the global balance condition [10–12]. The event-chain Monte Carlo (ECMC) algorithm, originally proposed for hard-sphere systems [13], is one of such algorithms breaking the detailed balance condition. This efficient algorithm enables us to simulate about 10^6 particles in equilibrium [14]. Furthermore, it can be generalized for more general particle systems such as soft-sphere and Lennard–Jones particles

[15, 16], and continuous spin systems such as XY and Heisenberg spin models [17, 18]. In the next section, we describe the ECMC algorithm for particle systems and continuous spin systems.

2. The event-chain Monte Carlo algorithm

Algorithm 1 describes the ECMC algorithm for hard-sphere systems in d dimensions. We denote the radius of particles as σ, the total number of particles as N, the locations of particles as $\{\boldsymbol{x}_k\}_{k=0,\cdots,N}$, and the standard basis of $\mathbb{R}^d$ as $\{\boldsymbol{e}_k\}_{k=0,\cdots,d-1}$. In the function DISPLACEMENT $(i, \boldsymbol{v})$, we compute how long i-th sphere can move along the vector $\boldsymbol{v}$ until it collides with another particle, and the collided particle starts to move along the same vector $\boldsymbol{v}$. Consequently, many particles are moved along the same vector $\boldsymbol{v}$ until the total displacement of particles reaches ℓ, which is a tuning parameter of this algorithm. The displacement until the collision is uniquely determined by the configuration and the vector $\boldsymbol{v}$, and thus the dynamics of particles is deterministic for given $\boldsymbol{v}$, the initial particle i, and ℓ. This algorithm breaks detailed balance because particles move along the same direction $\boldsymbol{v}$ and never go back to the former position [13].

For the hard-sphere systems we can define a collision as the time when the distance between the moving particle i and another particle is 2σ because the pair potential is 0 for non-overlapping configurations and infinity for otherwise. However, for general particle systems with an interaction potential such as Lennard–Jones systems, the pair potential for arbitrary distance always takes a finite value, and thus a collision cannot be defined in the same manner as the case of hard-sphere systems. In order to define a collision for more general systems and generalize the

Algorithm 1 ECMC for hard sphere systems in d dimensions

Input N, ℓ, $\{\boldsymbol{x}_0, \cdots, \boldsymbol{x}_{N-1}\}$, σ
for $t \geq 0$ **do**
  $(i, \boldsymbol{v}) \leftarrow$ RANDOM(N, d)
  $s \leftarrow 0$
  while $s < \ell$ **do**
    $(\delta, j) \leftarrow$ DISPLACEMENT$(i, \boldsymbol{v})$
    if $s + \delta < \ell$ **then**
      $(\boldsymbol{x}_i, s) \leftarrow (\boldsymbol{x}_i + \delta\boldsymbol{v}, s + \delta)$
      $i \leftarrow j$
    else
      $(\boldsymbol{x}_i, s) \leftarrow (\boldsymbol{x}_i + (\ell - s)\boldsymbol{v}, \ell)$
    end if
  end while
end for
function RANDOM(N, d)
  Sample i uniformly from $\{0, \cdots, N-1\}$
  Sample $\boldsymbol{v}$ uniformly from $\{\boldsymbol{e}_0, \cdots, \boldsymbol{e}_{d-1}\}$
  return $(i, \boldsymbol{v})$
end function
function DISPLACEMENT$(i, \boldsymbol{v})$
  for $k \in \{0, \cdots, N-1\} \setminus \{i\}$ **do**
    Compute δ_k such that $|\boldsymbol{x}_i + \delta_k\boldsymbol{v} - \boldsymbol{x}_k| = 2\sigma$
  end for
  $j \leftarrow \arg\min_k \delta_k$
  $\delta \leftarrow \delta_j$
  return (δ, j)
end function

ECMC algorithm to them, three essential concepts are introduced [16]; the factorized Metropolis probability, infinitesimal moves, and an event-driven Monte Carlo scheme [19]. Thanks to the factorization of the Metropolis transition probability, one can determine whether a proposal of a new state is accepted or not for each interacting pair independently, and the proposal is accepted only if all the interacting pairs accept it. If a new state which is infinitesimally close to the former state continues to be proposed, then at most a particle interacting with the moving one can reject it; the probability that more than two pairs simultaneously reject the proposal with infinitesimal displacements is higher-order infinitesimal. We define a collision as the probabilistic rejection which is caused by up to one interacting pair. Furthermore, an event-driven Monte Carlo scheme [19] allows us to compute the displacement until a collision efficiently. Consequently, a collision is defined in a probabilistic manner, and the ECMC algorithm is generalized for particle system with interaction potentials.

We present the ECMC algorithm for more general systems in Algorithm 2, where β is the inverse temperature. The Hamiltonian of the system considered in the algorithm is given by

$$H\left(\boldsymbol{x}_0, \cdots, \boldsymbol{x}_{N-1}\right) = \sum_{i<j} E_{ij}\left(\boldsymbol{x}_i, \boldsymbol{x}_j\right).$$

Note that $\boldsymbol{x}_i$ represents a location of i-th particle for particle systems, or components of i-th spin for continuous spin systems. In the algorithm, p is the number of degrees of freedom per one particle or one spin, and $\{\boldsymbol{v}_i\}_{i=0,\cdots,p-1}$ is a set of linearly independent vectors. The state $\boldsymbol{x}_i$ is updated by an operator $T_{\boldsymbol{v}}(s)$; for particle systems $T_{\boldsymbol{v}}(s)\,\boldsymbol{x}_i = \boldsymbol{x}_i + s\boldsymbol{v}$, and for continuous spin systems $T_{\boldsymbol{v}}(s)\,\boldsymbol{x}_i = R_{\boldsymbol{v}}(s)\,x_i$ where $R_{\boldsymbol{v}}(s)$ is a rotation matrix around the vector $\boldsymbol{v}$ with an angle s.

It is revealed that the ECMC algorithm outperforms other conventional algorithms in various systems [13, 15, 17, 20], and we can simulate very large systems consisting of about 10^6 particles or spins in equilibrium by using the algorithm [14, 21, 22]. One of successful examples of the algorithm is the three-dimensional ferromagnetic Heisenberg spin model, in which the algorithm reduces the value of the dynamical critical exponent z from $z = 2$ to $z \simeq 1$ [18].

3. Application to a three-dimensional classical Heisenberg spin model

We consider a classical Heisenberg spin model of a chiral helimagnet in three dimensions defined by the Hamiltonian

$$H\left(\{\boldsymbol{S}_i\}\right) = -J\sum_{\langle i,j\rangle} \boldsymbol{S}_i \cdot \boldsymbol{S}_j - \boldsymbol{D}\cdot\sum_i \left(\boldsymbol{S}_i \times \boldsymbol{S}_{i+\hat{y}}\right) - \boldsymbol{h}\cdot\sum_i \boldsymbol{S}_i, \tag{1}$$

where $\boldsymbol{S}_i$ is a Heisenberg spin with unit length, $J > 0$ is a ferromagnetic coupling constant of an ordinary exchange interaction, and $\boldsymbol{h} = h\hat{z}$ is a magnetic field. The bracket $\langle\cdot,\cdot\rangle$ in the first term of the Hamiltonian represents the neighboring pairs of sites, and the other summations run over all the sites. The second term in the Hamiltonian is a uniaxial Dzyaloshinskii–Moriya (DM) interaction with a DM vector $\boldsymbol{D} = D\hat{y}$, which induces a helical structure into the system. The ground state of the system with/without the magnetic field was studied by a variational analysis of the one-dimensional continuum model [23–25]. When $h = 0$, the ground state has a simple helical structure characterized by one wave vector $\boldsymbol{q}_{\mathrm{chiral}} = \arctan\left(D/J\right)\hat{y}$, and when $h > h_{\mathrm{c}}$ all of spins are parallel to the magnetic field $\boldsymbol{h}$. A nontrivial and interesting structure, the chiral magnetic soliton lattice (CSL) structure, emerges in the intermediate region $0 < h < h_{\mathrm{c}}$. The CSL structure is periodic as the helical structure at $h = 0$, but multiple wave vectors are needed to characterize the CSL structure.

We study the model at finite temperature with $D/J = 1$ in a three-dimensional simple cubic lattice. The lattice is a cuboid where the linear size of y direction is 8 times as long as x and z

Algorithm 2 ECMC for more general potentials

Input N, ℓ, $\{\boldsymbol{x}_0, \cdots, \boldsymbol{x}_{N-1}\}$, $\{E_{ij}(\boldsymbol{x}_i, \boldsymbol{x}_j)\}_{i,j=0,\cdots,N-1}$, β
for $t \geq 0$ **do**
    $(i, \boldsymbol{v}) \leftarrow$ RANDOM(N, p)
    $s \leftarrow 0$
    while $s < \ell$ **do**
        $(\delta, j) \leftarrow$ DISPLACEMENT$(i, \boldsymbol{v})$
        if $s + \delta < \ell$ **then**
            $(\boldsymbol{x}_i, s) \leftarrow (T_{\boldsymbol{v}}(\delta)\, \boldsymbol{x}_i, s + \delta)$
            $i \leftarrow j$
        else
            $(\boldsymbol{x}_i, s) \leftarrow (T_{\boldsymbol{v}}(\ell - s)\, \boldsymbol{x}_i, \ell)$
        end if
    end while
end for
function RANDOM(N, p)
    Sample i uniformly from $\{0, \cdots, N-1\}$
    Sample $\boldsymbol{v}$ uniformly from $\{\boldsymbol{v}_0, \cdots, \boldsymbol{v}_{p-1}\}$
    return $(i, \boldsymbol{v})$
end function
function DISPLACEMENT$(i, \boldsymbol{v})$
    for $k \in \{0, \cdots, N-1\} \setminus \{i\}$ **do**
        Sample r uniformly from $(0, 1]$
        Compute δ_k that satisfies

$$r = \exp\left(-\beta \int_0^{\delta_k} \max\left[0, \frac{\partial E_{ik}(T_{\boldsymbol{v}}(s)\, \boldsymbol{x}_i, \boldsymbol{x}_k)}{\partial s}\right] \mathrm{d}s\right)$$

    end for
    $j \leftarrow \arg\min_k \delta_k$
    $\delta \leftarrow \delta_j$
    return (δ, j)
end function

directions. The linear size of x and z directions of the lattice is denoted by L and the total number of sites is $N = 8L^3$. Periodic boundary conditions are imposed on x and z directions and free boundary condition on y direction. We perform large-scale equilibrium Monte Carlo simulations using the ECMC algorithm combined with the heat-bath algorithm, the over-relaxation updates [26, 27], and the exchange Monte Carlo method [7]. The largest system size in our simulations is $L = 64$, where the total number of spins $N = 64 \times 512 \times 64 > 10^6$.

We present in Fig. 1 the wave-number dependence of the magnetic susceptibility $\chi^{\parallel}(q)$ for $h/J = 0.1$, 0.2, and 0.3 at low temperatures. Here, the wave-number-dependent magnetization and the magnetic susceptibility with a wave-vector $\boldsymbol{q} = q\hat{y}$ parallel to the DM vector $\boldsymbol{D}$ is defined as

$$\begin{aligned} \boldsymbol{m}^{\parallel}(q) &= \frac{1}{N}\sum_i \boldsymbol{S}_i \exp(\mathrm{i}\boldsymbol{q}\cdot\boldsymbol{r}_i), \\ \chi^{\parallel}(q) &= N\beta\left(\left\langle\left|\boldsymbol{m}^{\parallel}(q)\right|^2\right\rangle - \left|\left\langle\boldsymbol{m}^{\parallel}(q)\right\rangle\right|^2\right). \end{aligned}$$

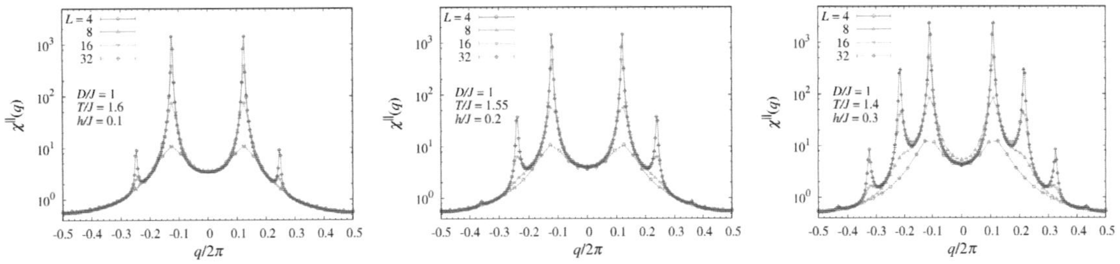

Figure 1. Wave-number dependence of $\chi^{\parallel}(q)$ of the chiral helimagnetic model in three dimensions for various system sizes in the low-temperature region. The values of the magnetic fields perpendicular to the DM vector are (left) $h/J = 0.1$, (center) $h/J = 0.2$, and (right) $h/J = 0.3$.

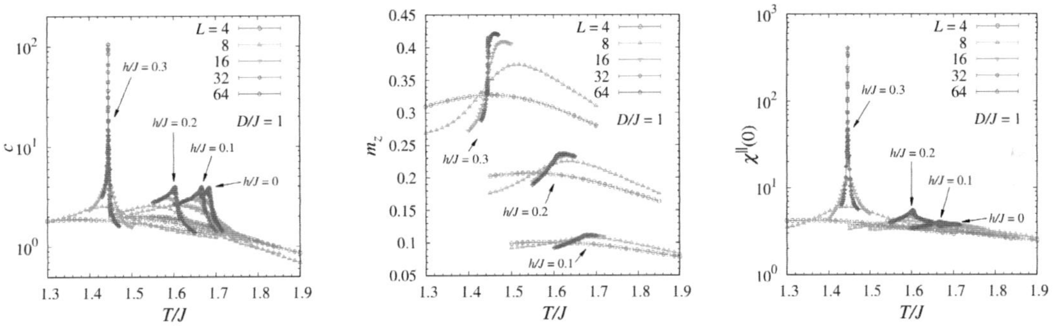

Figure 2. Temperature dependence of (left) the specific heat c, (center) the magnetization m_z parallel to the magnetic field $\boldsymbol{h}$, and (right) the uniform magnetic susceptibility χ of the three-dimensional chiral helimagnetic model.

One can see in the figure that there exist multiple peaks for larger system sizes which characterize the CSL structure. While the CSL structure emerges in the low-temperature region independent of the magnetic field h/J in our simulations, other thermodynamic quantities show completely different behaviors depending on the value of h/J. In Fig. 2, we show the specific heat $c = N\beta^2(\langle e^2\rangle - \langle e\rangle^2)$, the magnetization parallel to the magnetic field $m_z = \sum_i \hat{z} \cdot \boldsymbol{S}_i/N$, and the uniform magnetic susceptibility $\chi^{\parallel}(0)$ for $h/J = 0.1$, 0.2, and 0.3. The specific heat and the uniform magnetic susceptibility for $h/J = 0$ are also plotted. As shown in the figure, the specific heat of the system for $h/J = 0.3$ shows a strong tendency to diverge at its transition temperature while that for $h/J = 0$, 0.1, and 0.2 has a finite value at their transition temperatures. Note that the system without the magnetic field belongs to the same universality class of the three-dimensional XY model [22], and thus the specific heat does not diverge at the critical temperature in the thermodynamic limit. More strikingly, the uniform magnetic susceptibility for $h/J = 0.3$ also has a strong tendency to diverge at the finite temperature although the uniform magnetization does not characterize the CSL structure in the low-temperature region, and the magnetization parallel to the magnetic field changes rapidly near the transition temperature. These suggest that there exists a critical point $(T_{\rm d}, h_{\rm d})$ with $0.2 < h_{\rm d} < 0.3$ in the magnetic phase diagram of the system although the precise location of the critical point is not determined in this work. Recently, the existence of a critical point in the magnetic phase diagram of a magnetic compound $Cr_{1/3}NbS_2$, which has been extensively studied as a possible candidate of chiral helimagnets in experiments, is also suggested experimentally [28]. This might be consistent with that found in this work. In Ref. [25], Dzyaloshinskii predicts by analyzing a one-dimensional continuum model that the specific heat c diverges with a logarithmic correction as $c \sim \left((T_* - T)\log^2(T_* - T)\right)^{-1}$ from below the transition temperature T_* although it does not diverge from above the the transition temperature. The specific heat of the system with $h/J = 0.3$ in our results seems to also have an antisymmetric singular behavior. Larger-scale simulations of the system is needed to determine the critical exponents and examine the validity of Dzyaloshinskii's theory.

acknowledgement
The authors thank S. Hoshino and Y. Kato for very useful discussions. Numerical simulation in this work has mainly been performed by using the facility of the Supercomputer Center, Institute for Solid State Physics, the University of Tokyo. This research was supported by the Grants-in-Aid for Scientific Research from the JSPS, Japan (No. 25120010 and 25610102), and JSPS Core-to-Core program "Nonequilibrium dynamics of soft matter and information." This work was also supported by "Materials research by Information Integration Initiative (MI^2I)" project of the Support Program for Starting Up Innovation Hub from Japan Science and Technology Agency (JST).

References

[1] Metropolis N, Rosenbluth A W, Rosenbluth M N, Teller A H and Teller E 1953 *J. Chem. Phys.* **21** 1087
[2] Swendsen R H and Wang J S 1987 *Phys. Rev. Lett.* **58** 86
[3] Wolff U 1989 *Phys. Rev. Lett.* **62** 361
[4] Berg B A and Neuhaus T 1992 *Phys. Rev. Lett.* **68** 9
[5] Lyubartsev A P, Martsinovskii A A, Shevkunov S V and Vorontsov-Velyaminov P N 1992 *J. Chem. Phys.* **96** 1776
[6] Marinari E and Parisi G 1992 *Europhys. Lett.* **19** 451
[7] Hukushima K and Nemoto K 1996 *J. Phys. Soc. Jpn.* **65** 1604
[8] Wang F and Landau D P 2001 *Phys. Rev. Lett.* **86** 2050
[9] Iba Y 2001 *Int. J. Mod. Phys. C* **12** 623
[10] Suwa H and Todo S 2010 *Phys. Rev. Lett.* **105** 120603
[11] Turitsyn K S, Chertkov M and Vucelja M 2011 *Physica D* **240** 410
[12] Fernandes H C M and Weigel M 2011 *Comput. Phys. Commun.* **182** 1856
[13] Bernard E P, Krauth W and Wilson D B 2009 *Phys. Rev. E* **80** 056704
[14] Bernard E P and Krauth W 2011 *Phys. Rev. Lett.* **107** 155704
[15] Peters E A J F and de With G 2012 *Phys. Rev. E* **85** 026703
[16] Michel M, Kapfer S C and Krauth W 2014 *J. Chem. Phys.* **140** 054116
[17] Michel M, Mayer J and Krauth W 2015 *Europhys. Lett.* **112** 20003
[18] Nishikawa Y, Michel M, Krauth W and Hukushima K 2015 *Phys. Rev. E* **92** 063306
[19] Bortz A B, Kalos M H and Lebowitz J L 1975 *J. Comput. Phys.* **17** 10
[20] Isobe M and Krauth W 2015 *J. Chem. Phys.* **143** 084509
[21] Kapfer S C and Krauth W 2015 *Phys. Rev. Lett.* **114** 035702
[22] Nishikawa Y and Hukushima K 2016 (*Preprint* `arXiv:1603.04200`)
[23] Dzyaloshinskii I E 1964 *Sov. Phys. JETP* **19** 960
[24] Dzyaloshinskii I E 1964 *Sov. Phys. JETP* **20** 223
[25] Dzyaloshinskii I E 1965 *Sov. Phys. JETP* **20** 665
[26] Creutz M 1987 *Phys. Rev. D* **36** 515
[27] Brown F R and Woch T J 1987 *Phys. Rev. Lett.* **58** 2394
[28] Tsuruta K, Mito M, Deguchi H, Kishine J, Kousaka Y, Akimitsu J and Inoue K 2016 *Phys. Rev. B* **93** 104402

Magnitude of the Wang-Landau error

Gregory Brown[1], PA Rikvold[1], Kh Odbadrakh[2] and DM Nicholson[3]

[1] Department of Physics, Florida State University, Tallahassee, FL 32306, USA

[2] Joint Institute for Computational Sciences at ORNL, University of Tennessee, Oak Ridge, TN 37831, USA

[3] Department of Physics, University of North Carolina Asheville, Asheville, NC 28804, USA

E-mail: gbrown@fsu.edu

Abstract. The Wang-Landau algorithm is an entropic sampling method that incoroporates an update factor $\ln f_i$, which introduces a self-avoidance tendency into the random walk. Continued sampling at constant $\ln f_i$ leads to a steady state estimate of the density of states $\ln g_i(E)$. We find numerically that the difference between $\ln g_i(E)$ and the true density of states $\ln g(E)$ is proportional to the update factor.

In general, the name "flat-histogram sampling" is applied to importance sampling Monte Carlo methods in which each proposed step in a Markov chain is accepted with probability $p(E \to E') = \min[1, g(E)/g(E')]$, with $g(E)$ the density of states (DOS). One result is that

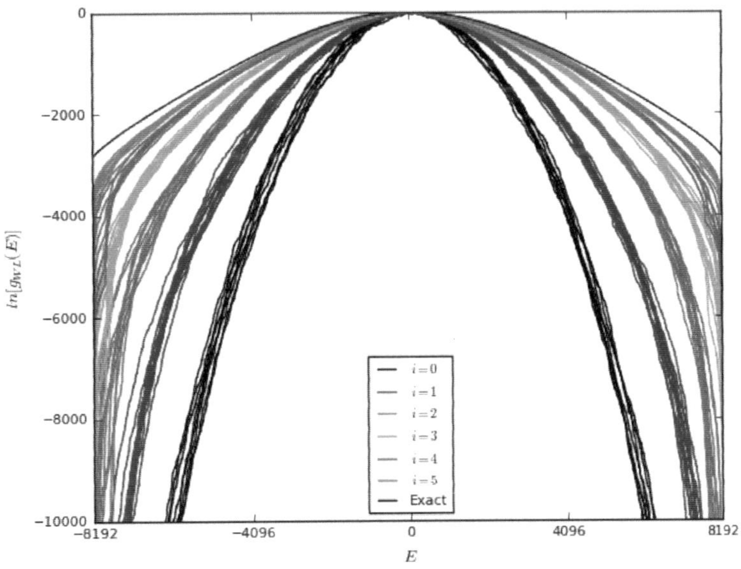

Figure 1. Snapshots of the estimated DOS $\ln[g_i(E)]$ for the 64×64 square-lattice Ising ferromagnet.

Journal of Physics: Conference Series **750** (2016) 012015 doi:10.1088/1742-6596/750/1/012015

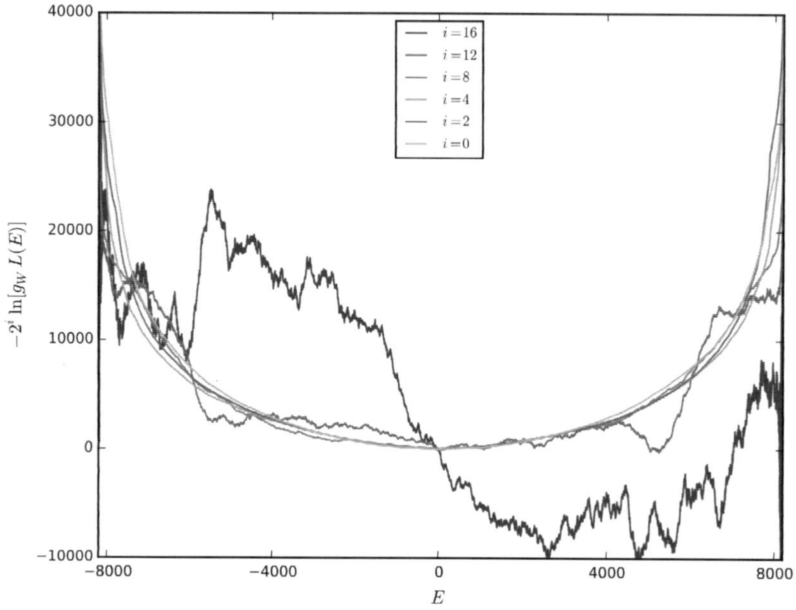

Figure 2. Wang-Landau error $g_{i,\mathrm{WL}}(E)$ scaled by the update factor $\ln f_i$ for several stages i in the algorithm applied to the 64×64 square-lattice Ising ferromagnet. Here $\ln g_{\mathrm{FIX}}(E)$ is the exact result from Ref. [5].

the simulation-time-averaged histogram is a constant density with respect to energy, $h(E) = C$. When the DOS is not known, the Wang-Landau algorithm [1] can be used to generate an estimate of the DOS, $g_i(E)$. Each iterative stage i of the algorithm is characterized by an update factor $\ln f_i$, which gets smaller as the stages progress. Typically $\ln f_i = 2^{-i}$ and the Wang-Landau estimated DOS becomes self-consistent as $\ln f_i \to 0$. Within each stage, the estimated DOS is dynamically updated after each step in the Markov chain by adding $\ln f_i$ to the logarithm of the estimated DOS at the current energy of the Monte Carlo walker, $\ln[g_i(E)] \leftarrow \ln[g_i(E)] + \ln f_i$. Within each stage i of the Wang-Landau algorithm the dynamic estimate of the function $\ln[g(E)]$ will approach a steady-state value [2, 3]. In this way, the Wang-Landau algorithm can be said to produce a series of estimated DOS g_i that converge to the true DOS g, particularly for the $1/t$ algorithm [3, 4].

The estimated DOS, $\ln[g_i(E)]$, for one run of the Wang-Landau algorithm is shown in Fig. 1 for the 64×64 square-lattice Ising ferromagnet. Several intermediate results are shown for each stage i and the different stages are color coded (online), but can be easily distinguished visually. It is clear from these data that $\ln[g_i(E)]$ approaches a steady-state value for each i, and that this estimate relaxes to $\ln[g(E)]$ as i increases. The exact DOS, as calculated using the result of Beale [5], is also shown.

In our implementation of the Wang-Landau algorithm, we have separated the estimated DOS at each stage into two factors

$$\ln[g_i(E)] = \ln[g_{\mathrm{FIX}}(E)] + \ln[g_{i,\mathrm{WL}}(E)] , \tag{1}$$

with $g_{\mathrm{FIX}}(E)$ fixed at the beginning of the Wang-Landau algorithm and unchanged throughout all stages. Then $g_{i,\mathrm{WL}}(E)$ isolates the dynamically changing part of the estimated DOS. In particular, if g_{FIX} is the true $g(E)$, then $\ln[g_{i,\mathrm{WL}}(E)] = \ln[g_i(E)] - \ln[g(E)]$ is the error in the

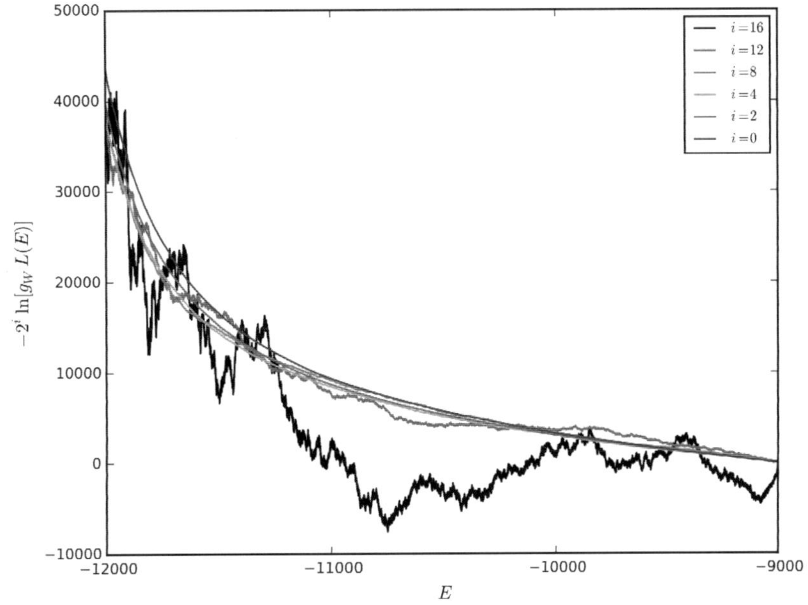

Figure 3. Wang-Landau error $g_{i,\mathrm{WL}}(E)$ scaled by the update factor $\ln f_i$ for several stages i in the algorithm applied to the $16 \times 16 \times 16$ Heisenberg ferromagnet on a simple cubic lattice. Here $\ln g_{\mathrm{FIX}}(E)$ was estimated from a previous, identical calculation.

stage-i Wang-Landau estimated DOS in the steady state limit. Using the exact result [5] as $g_{\mathrm{FIX}}(E)$, the error $g_{i,\mathrm{WL}}(E)$ is shown in Fig. 2. In this scheme the stages are run independently and the traditional flatness critierion is unimportant. Multiplying the DOS by the inverse update factor, $1/\ln f_i = 2^i$, results in good data collapse for a wide range of $\ln f_i$. As $\ln f_i$ get smaller, the noise in the error becomes more pronounced. For $i \geq 16$ the noise dominates. Similar behavior for the three-dimensional Heisenberg ferromagnet on a simple cubic lattice is shown in Fig. 3. For the Heisenberg ferromagnet, the energy has been restricted to the window $E \in [-12000, -9000]$ while the paramagnetic peak lies at $E = 0$, in part because the DOS decreases rapidly near the ground state. We conjecture that it is a general feature of the Wang-Landau algorithm for the systematic error to include the update factor $\ln f_i$ as a multiplicative factor.

Acknowledgments

GB and KO were supported by the Center for Defect Physics (CDP), an Energy Frontier Research Center funded by the U.S. Department of Energy, Office of Science, Basic Energy Sciences. PAR acknowledges support from NSF grant No. DMR-1104829. DMN was supported by the Materials Science an Engineering Division, Office of Basic Energy Sciences, US Department of Energy.

References

[1] Wang F and Landau DP 2001 *Phys. Rev. Lett.* **86** 2050; 2001 *Phys. Rev.* E **64** 056101
[2] Zhou C and Bhatt RN 2005 *Phys. Rev.* E **72** 025701(R)
[3] Belardinelli RE and Pereyra VD 2007 *J. Chem. Phys.* **127** 184105
[4] Zhou C and Su J *Phys. Rev.* E **78** 046705
[5] Beale PD *Phys. Rev. Lett.* **76** 78

Study of the Antiferromagnetic Blume-Capel Model on kagomé Lattice

Chi-Ok Hwang[1], **Sojeong Park**[1], **Wooseop Kwak**[2]

[1]Department of Physics, GIST, Gwangju, 500-712, South Korea

[2]Department of Physics, Chosun University, Gwangju, 501-759, South Korea

E-mail: wkwak@chosun.ac.kr

Abstract.

We study the anti-ferromagnetic (AF) Ising model and the AF Blume-Capel (BC) model on the kagomé lattice. Using the Wang-Landau sampling method, we estimate the joint density functions for both models on the lattice, and we obtain the exact critical magnetic fields at zero temperature by using the micro-canonical analysis. We also show the patterns of critical lines for the models from micro-canonical analysis.

1. Introduction

Monte Carlo simulations currently play a major role in statistical physics to study phase transitions and critical properties. They are well-known in the case of anti-ferromagnetic (AF) Ising on square lattice. However, the studies of frustrated systems such as triangular [1, 2, 3, 4, 5, 6, 7, 8, 9, 10, 11, 12, 13, 14, 15, 16, 17, 18] and kagomé lattices [19, 20] are also highlighted by the difficult task to understand the various phenomena. It has been reported that AF Ising models have three types of conjectured patterns of critical lines [21]. The Blume-Capel (BC) model [22, 23, 24, 25, 26, 27] has been studied to understand systems such as metamagents, ternary alloys, and multicomponent fluids. This spin-1 Ising model shows first-order transitions and second-order transition. In this paper, we study the critical lines and properties of the AF BC model as well as those of the AF Ising model on the kagomé lattice using the Wang-Landau sampling method [28, 29, 30, 31] and microcanonical analysis [1].

2. Model and Simulation

The Hamiltonian of the AF Ising model defined on $L \times L$ square lattice in two dimensions is given by

$$\mathcal{H} = -JE - hM, \tag{1}$$

where $E = \sum_{\langle i,j \rangle} \sigma_i \sigma_j$ and $M = \sum_i \sigma_i$, and $\sigma_i = \pm 1$. $J < 0$, and h is an external magnetic field. The Hamiltonian of the AF BC model is as follows:

$$\mathcal{H} = -JE - hM + DM^2, \tag{2}$$

where D is the crystal field (also called the single-spin anisotropy parameter or the spin impurity chemical potential), $M^2 = \sum_i {\sigma_i}^2$, and $\sigma_i = -1, 0,$ or 1. Here, a negative coupling constant

($J < 0$) defines the AF BC model. The physical origin of the crystal field D arises from the non-central potentials for the metal atoms coordinated with various ligands in the crystal.

Using the WL algorithm, the $3d$ random walks are performed in the joint space [32, 33] $A = (E, M)$ for the AF Ising model and $A = \left(E, M, M^2\right)$ for the AF BC model by randomly changing the states of spins A, where the order parameter M is $\sum_i S_i$ and the square of order parameter M^2 is $\sum_i {S_i}^2$, but the state A associated with each spin configuration is only accepted with a probability proportional to the reciprocal of the joint density of states $g\left(A\right)$ [32, 33, 34, 35]. Therefore, the transition probability from state A to A' is

$$p\left(A \rightarrow A'\right) = min\left(\frac{g(A)}{g(A')}, 1\right), \tag{3}$$

which indicates that if $g\left(A'\right) \leq g\left(A\right)$, a state with spin configuration A' is always accepted, and that if $g\left(A'\right) > g\left(A\right)$, it is accepted with the probability $g\left(A\right)/g\left(A'\right)$.

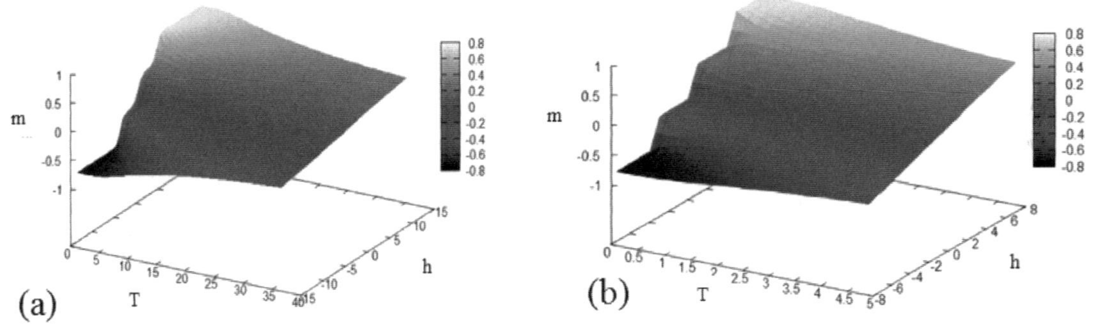

Figure 1. Magnetization m of (a) AF Ising model with $L = 12$ and (b) AF BC model with $L = 6$ as a function of energy, temperature, and h field on kagomé lattice.

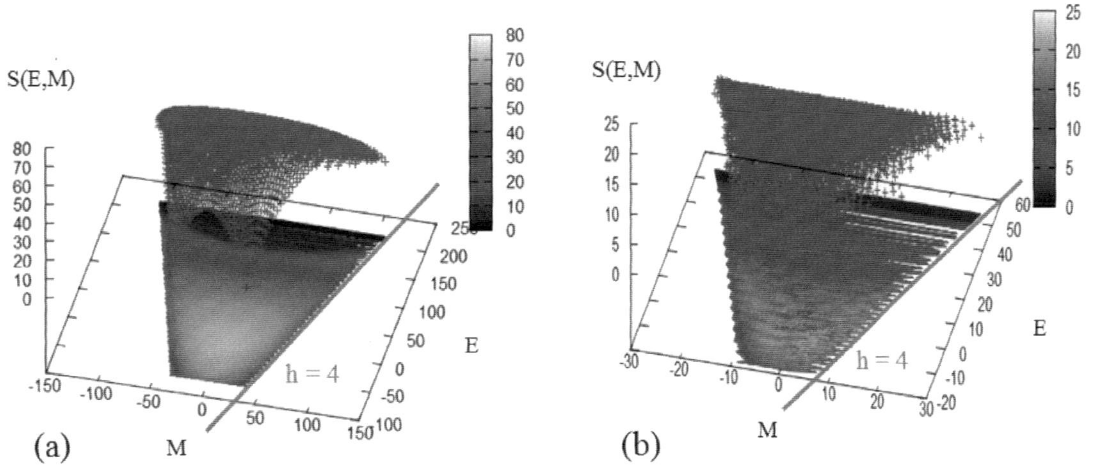

Figure 2. Microcanonical entropy of (a) AF Ising model with $L = 12$ and (b) AF BC model with $L = 6$ as a function of energy and magnetization on kagomé lattice.

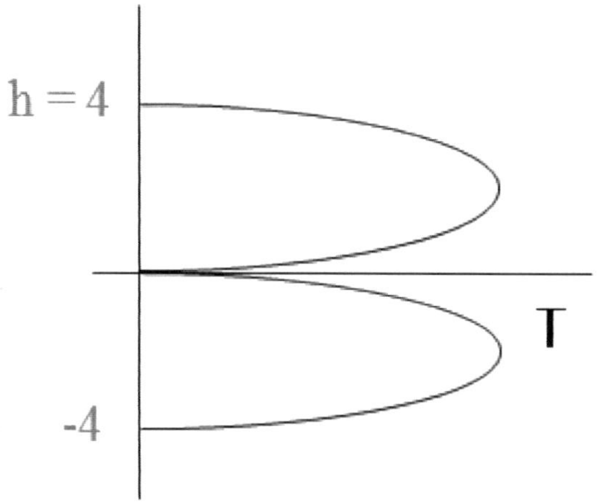

Figure 3. Schematic plot of three types of critical lines of a) AF Ising model with $L = 12$ and (b) AF BC model with $L = 6$ on kagomé lattice.

3. Results and Discussion

The canonical distributions of magnetization $m = M/L^2$ for AF Ising model with linear dimension $L = 12$ and AF BC model with $L = 6$ on on kagomé lattice are shown in figure 1. The line of m as a function of h in the vicinity of $T \sim 0$ shows discontinuity for AF BC model, but it does not show for AF Ising. Figure 2 shows microcanonical entropy $S(E, M)$ of (a) AF Ising model with $L = 12$ and (b) AF BC model with $L = 6$ as a function of energy and magnetization on kagomé lattice. In Figure 2, the contour plot of the ME diagram has four vertices. The top-left vertices corresponds to the state of all spins up. The locations of top-right vertices are as follows: $E = 216$ and $M = 108$ for AF Ising model, and $E = 54$ and $M = 27$ for AF BC model. In the contour plot of the ME diagram in Figure 2, the bottom line corresponds to the frustrated AF ground state without h. The locations of bottom-right vertices are $E = -72$ and $M = 36$ for AF Ising model, and $E = -18$ and $M = 9$ for AF BC model.

We can observe that at zero temperature the term with the minimum E_t is dominant, where $E_t = (E - hM)$ is the total energy. The total energy E_t can be interpreted as the intersection on the E axis of the linear line $E = hM + E_t$ in the ME diagram. Note that the slope of the line connecting the top right vertex and the bottom right vertex is six which is the critical magnetic field h_c, where $h_c = 4$ for AF Ising and AF BC models on kagomé lattice.

Figure 3 shows the estimated critical lines of the AF Ising and AF BC models on kagomé lattice. Although the patterns of critical lines for both models are the same, the transitions in the vicinity of $T \sim 0$ are different from each other. As shown in Figure 3, the critical line of AF BC model at $T \sim 0$ and $h = 4$ indicates the first order phase transition, whereas that of AF Ising at $T \sim 0$ and $h = 4$ indicates the second order phase transition.

4. Conclusions

In this preliminary study by using the small kagomé lattices, we study the AF BC and the AF Ising model on kagomé lattices using the Wang-Landau sampling method and micro-canonical anaysis. We find critical lines of both models and find the critical magnetic field $h_c = 4$ at zero temperature. We also find that the critical line of AF BC model at $T \sim 0$ indicates the first order phase transition and that of AF Ising indicates the second order phase transition. Thus, our future intention is to extend the present investigations to the critical properties and phenomena of the AF BC model on the kagomé lattice with nonzero D and non-zero h field.

Acknowledgment
This research was supported by Basic Science Research Program through the National Research Foundation of Korea (NRF) funded by the Ministry of Education Science and Technology (NRF-2013R1A1A2065043 for W.Kwak) and was also supported by the Research support projects for distinguished professors through a grant provided by GIST in 2016 (Grant No. GK03872 for C.-O.Hwang). In addition, this research has been performed as a collaborative research project of project No (Supercomputing infrastructure service and application for C.-O.Hwang) and supported by the KOREA INSTITUTE of SCIENCE and TECHNOLOGY INFORMATION (KISTI).

[1] C.-O. Hwang, S.-Y. Kim, D. Kang, J. M. Kim 2007 J.Stat.Mech. L05001
[2] C.-O. Hwang, S.-Y. Kim, D. Kang, J. M. Kim 2008 J. Korean Phys. Soc. **52** S203
[3] X. Yao 2010 Solid State Commun. **150** 160
[4] J.D. Noh, H. Rieger, M. Enderle, K. Knorr 2002 Phys. Rev. E **66** 026111
[5] Henk W. J. Blote, M. Peter Nightingale 1993 Phys. Rev. B **47** 15046
[6] C.-O. Hwang, S.-Y. Kim 2010 Physica A **389** 5650
[7] M. Zukovic, A. Bobak 2013 Phys. Rev. E **87** 032121
[8] O. Nagai, S. Miyashita, T. Horiguchi 1993 Phys. Rev. B **47** 202
[9] W. Li, S.-S. Gong, Y. Zhao, S.-J. Ran, S. Gao, G. Su 2010 Phys. Rev. B **82** 134434
[10] Y. L. Loh, D. X. Yao, E. W. Carlson 2008 Phys. Rev. B **77** 134402
[11] J.D. Noh, D. Kim 1992 Int. J. Mol. Sci. **6** 2913
[12] S.-H. Tsai, F. Wang, D.P Landau 2006 Braz. J. Phys. **36** 635
[13] M.A. Novotny, D.P. Landau 1981 Phys. Rev. B **24** 1468
[14] J.B. Santos-Filho, J.A. Plascak, D.P. Landau 2010 Physica A **389** 2934
[15] X. Qian, M. Wegewijs, Henk W. J. Blote 2004 Phys. Rev. E **69** 036127
[16] F.W.S. Lima, J.A. Plascak 2013 Eur. Phys. J. B **86** 1
[17] U. Yu 2015 Phys. Rev. E **91** 062121
[18] S. Hu, S.-H. Tsai, D.P. Landau 2014 Phys. Rev. E **89** 032118
[19] I. Syozi 1951 Progr. Theoret. Phys. **VI** 306
[20] K. KANO, S. NAYA 1953 Progr. Theoret. Phys. **10** 158
[21] L.-H. Gwa 1990 Phys. Rev. B **41** 7315
[22] M. Blume 1966 Phys. Rev. **141** 517
[23] H.W. Capel 1966 Physica (Amsterdam) **32** 966
[24] M. Blume, V.J. Emery, R.B. Griffiths 1971 Phys. Rev. A **4** 1071
[25] A.K. Jain, D.P. Landau 1980 Phys. Rev. B **22** 22.
[26] C.-J. Liu, H.-B. Sch'uttler 2002 Phys. Rev. E. **65** 056103
[27] W. Kwak, J. Jeong, J. Lee, D.H. Kim 2015 Phys. Rev. E. **92** 022134
[28] F. Wang, D. P. Landau 2011 Phys. Rev. Lett. **86** 2050
[29] D. P. Landau, S.-H. Tsai, M. Exler 2004 Am. J. Phys. **72** 1294
[30] B. J. Schulz, K. Binder, M. Müller, D. P. Landau 2003 Phys. Rev. E **67** 067102
[31] C. Zhou, T. C. Schulthess, S. Torbrügge, D. P. Landau 2006 Phys. Rev. Lett. **96** 120201
[32] D.P. Landau, F. Wang 2002 Computer Physics. Commun. **147** 674
[33] S.-H. Tsai, F. Wang, D.P. Landau 2007 Phys. Rev. E. **75** 061108
[34] S. Ryu, W. Kwak 2013 J. Korean Phys. Soc. **62** 559
[35] S. Ryu, W. Kwak 2013 J. Korean Phys. Soc. **62** 861

Journal of Physics: Conference Series **750** (2016) 012017 doi:10.1088/1742-6596/750/1/012017

Finite-size scaling of Lennard-Jones droplet formation at fixed density

Johannes Zierenberg and Wolfhard Janke

Institut für Theoretische Physik, Universität Leipzig, Postfach 100 920, 04009 Leipzig, Germany

E-mail: johannes.zierenberg@itp.uni-leipzig.de, wolfhard.janke@itp.uni-leipzig.de

Abstract. We reaccess the droplet condensation-evaporation transition of a three-dimensional Lennard-Jones system upon a temperature change. With the help of parallel multicanonical simulations we obtain precise estimates of the transition temperature and the width of the transition for systems with up to 2048 particles. This allows us to supplement previous observations of finite-size scaling regimes with a clearer picture also for the case of a continuous particle model.

1. Introduction

Despite the long-lasting scientific interest in droplet formation, it is still a modern problem with many open questions. This is partially due to the general nature of droplet formation, with relevance ranging from metastable decay over cloud formation to cluster formation in protein solutions. In this work, we consider equilibrium droplet formation which yields a firm basis to study the transition between a homogeneous gas and a mixed phase of a droplet in equilibrium with surrounding vapor [1, 2, 3, 4, 5]. The problem is formulated in the canonical ensemble. A common control parameter is the density while the temperature is fixed. This allows one, in principle, to access temperature-dependent quantities such as the isothermal compressibility and the interface tension.

At fixed temperature the theory has been verified by several computational studies, including lattice [6, 7, 8] and Lennard-Jones [9, 10, 11] systems. Instead, however, one can fix the density and vary the temperature which yields an equivalent but "orthogonal" finite-size scaling behavior [12, 13]. In this case, we recently observed an intermediate scaling regime consistent with finite-size scaling results for polymer aggregation [14]. Here, we present new results for the three-dimensional Lennard-Jones system at fixed density which extend and supplement our previous results [13].

2. Model and Method

We consider N Lennard-Jones particles in a three-dimensional box of length L with periodic boundary conditions, see Fig. 1. The self-avoidance and short-range attraction is modeled by the pairwise Lennard-Jones potential

$$V_{\rm LJ}(r_{ij}) = 4\epsilon\left[\left(\frac{\sigma}{r_{ij}}\right)^{12} - \left(\frac{\sigma}{r_{ij}}\right)^{6}\right], \tag{1}$$

Journal of Physics: Conference Series **750** (2016) 012017 doi:10.1088/1742-6596/750/1/012017

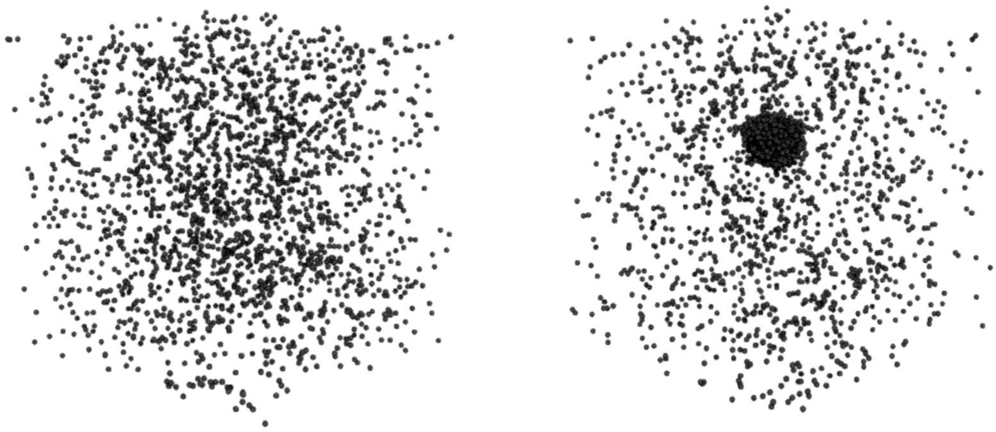

Figure 1. Illustration of a three-dimensional Lennard-Jones gas (left) and a droplet in equilibrium with surrounding vapor (right). Shown is a system with $N = 2048$ particles at density $\rho = 10^{-2}$ in a box with periodic boundary conditions.

where r_{ij} is the distance between particle i and j. As in Ref. [13], we set $\epsilon = 1$ and $\sigma = 2^{-1/6}$. The computational demand can be reduced by introducing a cutoff radius $r_c = 2.5\sigma$ above which particles do not interact anymore. The potential is then shifted by $V_{\mathrm{LJ}}(r_c)$ in order to be continuous, yielding

$$V^*_{\mathrm{LJ}}(r) = \begin{cases} V_{\mathrm{LJ}}(r) - V_{\mathrm{LJ}}(r_c) & r < r_c \\ 0 & \text{else} \end{cases} . \tag{2}$$

This is in accordance with the existing literature and enables the application of a domain decomposition, where the periodic box is decomposed into equally large (cubic) domains. These domains have to be at least of the size r_c. Then, the interaction of each particle is obtained by evaluating only its domain and the adjacent ones (in three dimensions this adds up to $3^3 = 27$ domains). Especially in the gas phase the simulation benefits from this procedure, where the particles are equally distributed in the full box.

Following Ref. [13], we fix the density $\rho = N/L^3$ and vary the temperature T. This allows the application of multicanonical simulations [15, 16, 17, 18], which are well-suited for first-order phase transitions such as condensation. At a first-order transition two phases are in coexistence with suppressed transition states in between. This is circumvented by replacing the canonical Boltzmann weight $\exp(-E/k_{\mathrm{B}}T)$ with an alterable weight function, which is iteratively adapted in order to yield a flat histogram in the energy. Each iteration is in equilibrium, sampling the distribution according to the current weight function. This leads to a straightforward parallelization [19, 20] which has been shown to perform very well for the problem at hand [8]. In the end, canonical expectation values are estimated by reweighting the data from a multicanonical production run.

A crucial aspect is the selection of Monte Carlo updates. While in Ref. [13] we restricted ourselves to local particles shifts, we added here particle "jumps" with a larger update range. The update range for a particle shift was set to $r = 0.3$ and for a particle jump to $r = L/2$. A new position is then selected with equal probability from a sphere of size r around the old position. This simple non-local update significantly enhanced the sampling of the gaseous phase and increased the performance of the simulation drastically. Instead of 512 we are now able to sample up to 2048 particles.

3. Theory

Here we only want to briefly recapture the results of many previous works [2, 3, 4, 5, 13]. For a supersaturated gas, it was shown that the probability of intermediate-sized droplets vanishes [3, 4], which leaves us with the scenario of a homogeneous gas phase on the one side and a droplet in equilibrium with surrounding vapor on the other side, see Fig. 1. At a fixed temperature, the droplet formation may be achieved by adding more particle excess to the already supersaturated gas. For both sides it is possible to formulate a contribution to the free-energy in terms of fluctuations (entropy of the gaseous phase) and surface tension (energy of the droplet). Importantly, the finite-size dependence could be rewritten in terms of the fraction of particle excess in the droplet as a function of a dimensionless density. This in turn allowed us to expand the results around the infinite-size transition temperature T_0 to yield the leading finite-size scaling behavior of the condensation temperature T_c at fixed density [13]. Similarly, we showed that an expansion of the free-energy difference at the condensation transition yields an estimate of the leading-order finite-size scaling of the transition rounding ΔT, i.e., the width of the transition. For details we refer to the prior literature and give here only the three-dimensional results to leading order:

$$T_c - T_0 \propto N^{-1/4}, \tag{3}$$

$$\Delta T \propto N^{-3/4}. \tag{4}$$

A crucial observation is now that the size R of the droplet at the condensation transition itself scales non-trivially with system size, namely in three dimensions $R \propto N^{1/4}$. Thus, the leading-order scaling may be identified in terms of powers of the linear extension of the droplet itself, $\propto R^{-1}$ respectively $\propto R^{-3}$. This is consistent with the interpretation that the droplet size at transition is the relevant system size. Then, a virtual subsystem around the droplet would lead to a transition between a homogeneous liquid (droplet) phase to a homogeneous gas phase with open boundary conditions. The competition between finite-size contributions from volume ($\propto R^3$) and surface ($\propto R^2$) would in this case give rise to an intuitive finite-size correction of the order R^{-1} [21, 22, 23, 13, 14].

4. Results

Due to the non-local Monte Carlo update, we are now able to extend our previous results [13] for the three-dimensional Lennard-Jones system to $N = 2048$ particles with improved statistics. The temperature scale is fixed by setting $k_\mathrm{B} = 1$. The finite-size transition temperature T_c is obtained as the location of the largest peak in the specific heat $C_V = k_\mathrm{B}\beta^2 \left(\langle E^2\rangle - \langle E\rangle^2\right)/N$ and plotted in Fig. 2 (left) versus the expected scaling behavior Eq. (3). Overall, the data qualitatively shows a linear behavior as predicted. A leading-order fit for $N \geq 1024$ yields $T_0 = 0.7032(3)$ with goodness-of-fit parameter $Q \approx 0.1$, shown in the figure as dashed black line. Including additional higher-order corrections, i.e., $T_c = T_0 + aN^{-1/4} + bN^{-2/4}$, yields $T_0 = 0.6979(2)$ for $N \geq 64$ with $Q \approx 0.1$. Both estimates differ from our previous results outside the error bars, which is expected because the fit error underestimates systematic uncertainties. This is partially due to the fact that we are not yet fully in the asymptotic scaling regime, which we will discuss below. The general trend of the results is compatible with our previous conclusion, where we already noted that the available three-dimensional Lennard-Jones system sizes were too small for clear results. However, it is worth noting that the current estimates are closer to each other, implying an infinite-size limit of the transition temperature around $T_0 \approx 0.70$.

The distance from the asymptotic scaling regime is best visible in the rounding of the transition. This is here obtained from the half width of the specific-heat peak, i.e., the width for which $C_V(T) > C_V^{\max}/2$, shown in Fig. 2 (right). Since the width vanishes in the thermodynamic

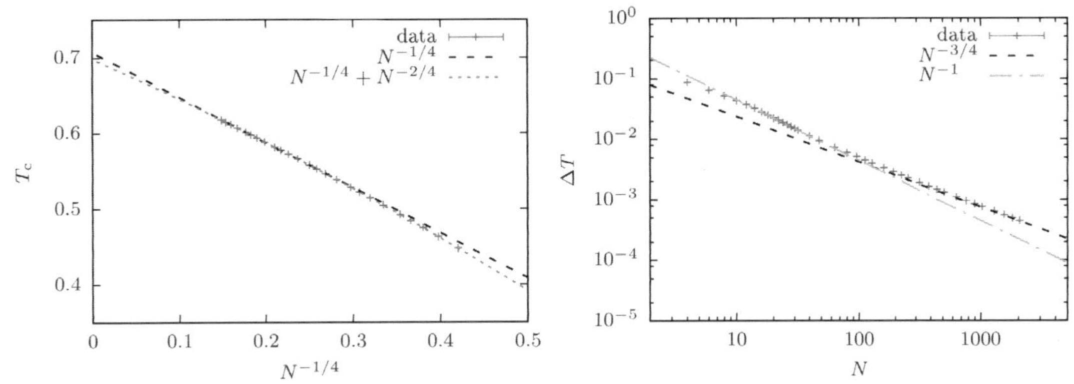

Figure 2. Finite-size scaling of the transition temperature (left) and rounding (right) as the maximum location and the half width of the specific-heat peak, respectively. Error bars are included but smaller than the symbol size.

limit, a double-logarithmic plot reveals the power-law scaling Eq. (4) as a straight line, here dashed black. As previously noticed [13], an intermediate scaling regime $\propto N^{-1}$ is clearly visible and marked with a dashed-dotted gray line. This is consistent with the droplet as relevant system size as in this regime still a large fraction of particles contribute to the transition droplet. However, for $N \geq 500$ it appears that the asymptotic scaling behavior slowly manifests itself and leading-order estimates become more reliable.

This can be tested by performing a direct fit of the power-law ansatz $\Delta T \propto N^{-\alpha}$, shown in Fig. 3 for variable lower bound N_{min}. Only for $N_{\mathrm{min}} \geq 640$ the fit yields plausible goodness-of-fit values $Q > 0.1$. Still, the distance to the predicted asymptotic scaling regime is noticeable. This is consistent with results obtained for the three-dimensional lattice gas where decent results were only obtained for $N_{\mathrm{min}} \approx 2000$ [13]. This emphasizes once more that care has to be taken with leading-order finite-size scaling away from the asymptotic scaling regime. While estimates of the thermodynamic limit become better with increasing system size, fit errors are difficult to judge because the underlying ansatz is not accurate enough. The same holds true for additional higher-order corrections. While the results are surely good estimates, the fit errors are a difficult measure, not capturing the uncertainty that comes from an incomplete fit model [24].

5. Conclusions

We have extended our previous results [13] for the three-dimensional Lennard-Jones system to larger system sizes. This allows for more consistent estimates of the thermodynamic limit when considering leading-order and higher-order fits. Still, the current Lennard-Jones system with up to 2048 particles at density $\rho = 10^{-2}$ is quite far away from the asymptotic scaling limit. We notice that the rounding of the transition is a good indicator to visualize this distance from the asymptotic scaling regime. Here, also the emerging intermediate scaling regime is best noticeable.

Acknowledgments

The project was funded by the European Union, the Free State of Saxony and the Deutsche Forschungsgemeinschaft (DFG) under Grant No. JA 483/31-1. The authors gratefully acknowledge the computing time provided by the John von Neumann Institute for Computing (NIC) on the supercomputer JURECA at Jülich Supercomputing Centre (JSC) under Grant No. HLZ24. Part of this work has been financially supported by the DFG through the Leipzig

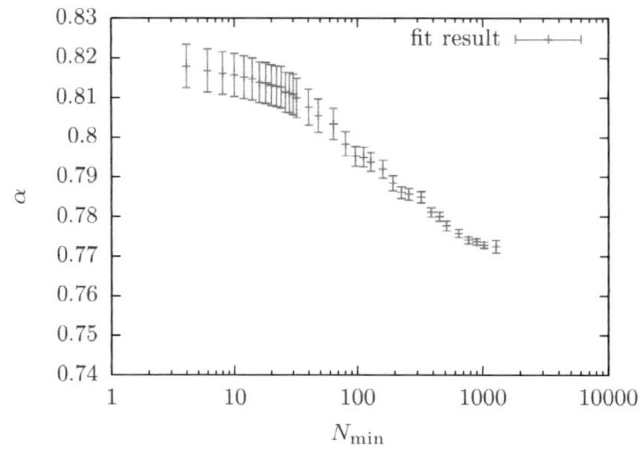

Figure 3. Results of direct fits of the ansatz $\Delta T \propto N^{-\alpha}$ to the data in Fig. 2 (right) with variable lower bound $N_{\rm min}$. The dotted line shows the predicted asymptotic scaling exponent.

Graduate School of Natural Sciences "BuildMoNa" and by the Deutsch-Französische Hochschule (DFH-UFA) through the Doctoral College "$\mathbb{L}^4$" under Grant No. CDFA-02-07.

References

[1] Binder K and Kalos M H 1980 *J. Stat. Phys.* **22** 363
[2] Neuhaus T and Hager J 2003 *J. Stat. Phys.* **113** 47
[3] Biskup M, Chayes L and Kotecký R 2002 *Europhys. Lett.* **60** 32
[4] Biskup M, Chayes L and Kotecký R 2003 *Commun. Math. Phys.* **242** 137
[5] Binder K 2003 *Physica A* **319** 99
[6] Nußbaumer A, Bittner E, Neuhaus T and Janke W 2006 *Europhys. Lett.* **75** 716
[7] Nußbaumer A, Bittner E and Janke W 2008 *Phys. Rev. E* **77** 041109
[8] Zierenberg J, Wiedenmann M and Janke W 2014 *J. Phys.: Conf. Ser.* **510** 012017
[9] MacDowell L G, Virnau P, Müller M and Binder K 2004 *J. Chem. Phys.* **120** 5293
[10] MacDowell L G, Shen V K and Errington J R 2006 *J. Chem. Phys.* **125** 034705
[11] Schrader M, Virnau P and Binder K 2009 *Phys. Rev. E* **79** 061104
[12] Martinos S, Malakis A and Hadjiagapiou I 2007 *Physica A* **384** 368
[13] Zierenberg J and Janke W 2015 *Phys. Rev. E* **92** 012134
[14] Zierenberg J, Mueller M, Schierz P, Marenz M and Janke W 2014 *J. Chem. Phys.* **141** 114908
[15] Berg B A and Neuhaus T 1991 *Phys. Lett. B* **267** 249
[16] Berg B A and Neuhaus T 1992 *Phys. Rev. Lett.* **68** 9
[17] Janke W 1992 *Int. J. Mod. Phys. C* **3** 1137
[18] Janke W 1998 *Physica A* **254** 164
[19] Zierenberg J, Marenz M and Janke W 2013 *Comput. Phys. Commun.* **184** 1155
[20] Zierenberg J, Marenz M and Janke W 2014 *Physics Procedia* **53** 55
[21] Privman V and Rudnick J 1990 *J. Stat. Phys.* **60** 551
[22] Borgs C and Kotecký R 1995 *J. Stat. Phys.* **79** 43
[23] Borgs C, Kotecký R and Medved I 2002 *J. Stat. Phys.* **109** 67
[24] Young P 2015 *Everything You Wanted to Know About Data Analysis and Fitting but Were Afraid to Ask, SpringerBriefs in Physics* (Springer International Publishing) [arXiv:1210.3781 (2012)]

Journal of Physics: Conference Series **750** (2016) 012018 doi:10.1088/1742-6596/750/1/012018

Drag force on two disks moving in a granular bed

A Kuwabara[1], N Yoshioka[2], T Shimada[1], and N Ito[1,2]

[1] Department of applied physics, graduate school of engineering, the university of tokyo, 7-3-1 hongo, bunkyo-ku, tokyo 113-8656, japan
[2] Riken aics, 7-1-26 minatojima-minami-machi, chuo-ku, kobe, hyogo 650-0047, japan

E-mail: kuwabara@serow.t.u-tokyo.ac.jp

E-mail: naoki.yoshioka@riken.jp

E-mail: shimada@ap.t.u-tokyo.ac.jp

E-mail: ito@ap.t.u-tokyo.ac.jp

Abstract. The drag forces acting on a disk and two disks dragged at constant speeds in a two-dimensional granular bed are measured by an event-driven molecular dynamics simulation. The normal drag force on a disk in case of two disks are dragged in parallel , F_{2x}, is almost the same with the drag force on a disk in case of a single disk is dragged , F_{1x}, when they are apart enough. As the distance between the disks D decreases, F_{2x} increases until it has the maximum at a certain point, $D*$, and after that it decreases. When D is small enough, whether F_{2x} is larger than F_{1x} or not depends on the ratio of the average radius of the granular prticles to the one of the dragged objects.

1. Introduction

To know the drag force on objects in a granular bed is not only interesting in terms of rheology but also important in terms of industrial application. There have been many researches about the drag force on a object in a granular bed [1–4], but there are few researches about the drag force on multiple objects in a granular bed. The research related to the drag force when multiple objects move in a granular bed is conducted by J. M. Solano-Altamirano et al [5]. They examined the interaction force in the horizontal direction when two intruders are falling simultaneously into a granular bed, but didn't examined the drag force in the moving direction in detail. So in our research, we examined the drag force on a disk in the moving direction when two disks move in parallel and compared it with the one when a single disk moves in a two-dimensional granular bed by simulation.

2. Model

The simulation snapshots are shown in Fig. 1. At first the dragged objects is fixed as the midpoint of them (its center if there is only a single dragged object) is at the midpoint of two walls and their interval is D in y direction, and the granular particles which have random velocities are placed at random positions. Then two walls are approached to each other at constant speeds until the packing fraction of granular particles become ϕ_0, and they are stopped. After that granular particles are given random velocities and the systems are made uniform by the elastic collisions between the granular particles. Finally the granular particles are stopped and the objects are dragged at constant speeds.

Journal of Physics: Conference Series **750** (2016) 012018 doi:10.1088/1742-6596/750/1/012018

2.1. granular bed

The granular particles are hard disks whose radius is $r(1.0 \pm 0.07)$ (uniform distribution). The normal coefficient of restitution is 0.9 and there is no tangential friction. In general it is known that there may be infinite collisions within a finite time in the system where the energies are lost when hard disks collide. So in this simulation, we avoid this by making collisions elastic if the relative speed in normal direction is less than $v_c = 10^{-4}$. v_c, which is the parameter that is arbitrarily introduced, should be small enough not to affect the drag force. The drag force are measured for various v_c and it is made sure that the drag force doesn't change so much if v_c is small enough.

2.2. dragged object

The dragged objects are hard disks whose radii are 1.0. They collide with granular particles elastically. They are dragged at constant speeds $V_x = 1.0$ in x direction.

2.3. boundary condition

There are periodic boundary in x direction and fixed walls in y direction. The system size is $2000r$ in x direction and $500r$ in y direction. The system size is large enough that there is no effect of the boundary condition.

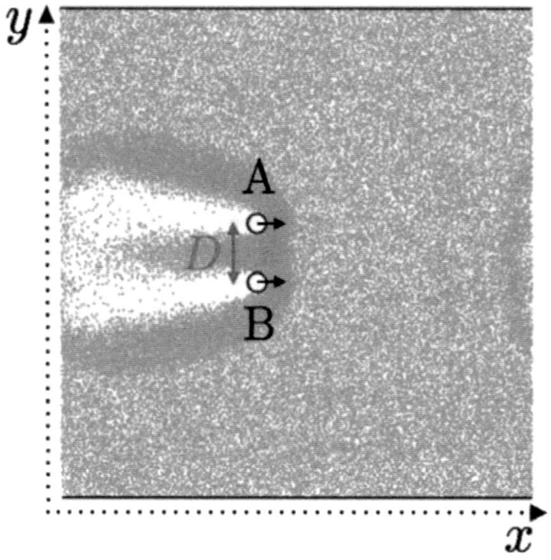

Fig. 1: The schematic picture of the simulation in case of dragging two object.

3. Results and Discussions

In this paper packing fraction ϕ_0 is fixed to 0.4. Drag force are averaged in the steady region and furthermore averaged over 10 samples. Drag force on a dragged object in case of only a disk is dragged is expressed as $\boldsymbol{F_1} = -F_{1x}\hat{\boldsymbol{e_x}} - F_{1y}\hat{\boldsymbol{e_y}}$. In case of two disks are dragged, the drag force on each object in x direction is considered to be the same, so drag force on one object is measured. Drag force on a dragged object A in case of there are two dragged objects is expressed as follows, $\boldsymbol{F_2} = -F_{2x}\hat{\boldsymbol{e_x}} - F_{2y}\hat{\boldsymbol{e_y}}$.

3.1. In case of only a single disk is dragged

Fig. 2(a) shows the snapshot of the simulation in case of only a disk is dragged. As can be seen from the figure, we can find that the granular particles are swept up by the dragged object and there are dense regions, sparse regions and regions in which there are almost no particles. To determine the boundary between the dense region and other regions, local packing fraction is calculated, and ϕ_{max} and ϕ_{th} are defined. ϕ_{max} is defined as the maximum value of ϕ_l in the region where $x \geq 0$ and the threshold of local packing fraction ϕ_{th} is defined as follows,

$$\phi_{th} = \frac{\phi_{max} + \phi_0}{2}. \tag{1}$$

We define the boundary as the coordinate which satisfies the equation $\phi_l(x, y) = \phi_{th}$. As can be seen form Fig. 2(a), this definition seems to be reasonable. We can find a boundary inside and outside the dense region. We call the outer boundary as the interface in this paper. Fig. 2(b) shows the interface for various r. The interface is less extended to the outside when r is small. It is considered as follows. If r is small the granular particles are more likely to collide with each other and lose energy. So granular particles are less likely to extend to the outside when they are smaller.

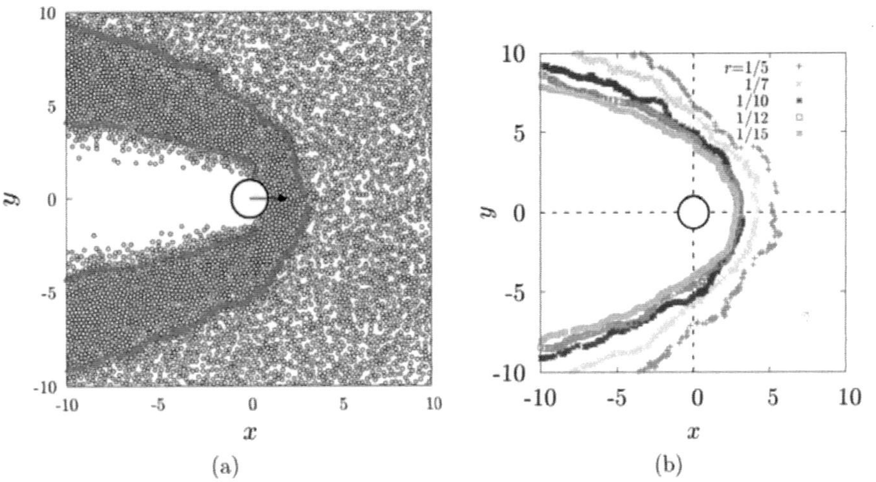

Fig. 2: (a): The snapshot of the simulation when $r = 1/10$. $\triangle$ denotes the coordinates of the interface where local packing fraction ϕ_l equals $\phi_{th}(= \frac{\phi_{max}+\phi_0}{2})$. (b): The shape of the interface for various r.

3.2. In case of two disks are dragged

Fig. 3(a) shows F_{2x}/F_{1x} as a function of D for various r. F_{2x} is found to increase as the distance between the disks D increases until it has the maximum at a certain point, $D*$, and then relaxes to F_{1x} when D is large enough. Fig. 3(b) shows the snapshot of the simulation. When D is large enough, we can find the two interfaces in front of the dragged objects and they don't interact with each other so much, so $F_{2x} \sim F_{1x}$. As can be seen from the figure, the distance D when two interfaces start to interact increases as r decreases. This is because the interface is more likely to be extended as r decreases (See Fig. 2(b)). As two dragged objects approaches two interfaces start to interact with each other and it becomes more difficult for granular particles to move through the objects, so drag force becomes larger. However if D is smaller than $D*$, it becomes easier for granular particles to move outside the dragged objects, so drag force becomes smaller. When two dragged objects are close enough, whether F_{2x} is smaller than F_{1x} or not depends on r. Basically, if the two dragged objects are close enough, drag force are assumed to be smaller when they are apart enough because the area where granular particles can collide with the dragged object become smaller than the one when they are apart enough. However if r is smaller the granular particles are more likely to be stuck just in front of the two dragged objects and granular particles in the region are hardly to move outside the two dragged objects. So drag force are considered to become larger when they are apart enough because of the frequent collision.

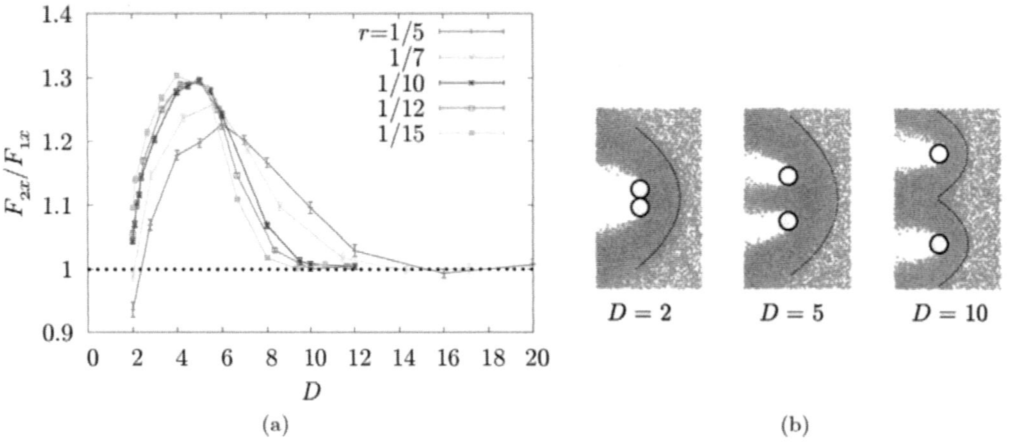

Fig. 3: (a): D dependence of F_{2x} for various r. (b): The simulation snapshot where $r = 1/10$.

4. Summary and Conclusion

We examined the drag forces acting on a disk and two disks dragged at constant speeds in a two-dimensional granular bed altering the average radius of the granular particles r. In case of a single disk is dragged, the granular particles are swept up by the dragged object and there are dense regions, sparse regions and regions in which there are almost no particles. The range of dense regions in front of the dragged object is wider when r is larger because they are less likely to collide than when r is smaller. In case of dragging two objects, The normal drag force on a disk in case of two disks are dragged in parallel, F_{2x}, is almost the same with the drag force on a disk in case of a single disk is dragged, F_{1x}, when they are apart enough. As the distance between the disks D decreases, F_{2x} increases until it has the maximum at a certain point, $D*$, and after that it decreases. When D is small enough, F_{2x} is smaller than F_{1x} if r is large. However F_{2x} is larger than F_{1x} if r is small. This reason is considered to be that if r is smaller the granular particles are more likely to be stuck just in front of the two dragged objects and granular particles in the region are hardly to move outside the two dragged objects, so drag force become larger because of the frequent collision.

References

[1] Wassgren C R, Cordova J a, Zenit R and Karion a 2003 *Phys. Fluids* **15** 3318–3330
[2] Chehata D, Zenit R and Wassgren C R 2003 *Phys. Fluids* **15** 1622
[3] Albert R, Pfeifer M, a L Barabási and Schiffer P 1999 *Phys. Rev. Lett.* **82** 205–208
[4] Takehara Y and Okumura K 2014 *Phys. Rev. Lett.* **112** 1–5
[5] Solano-Altamirano J M, Caballero-Robledo G a, Pacheco-Vázquez F, Kamphorst V and Ruiz-Suárez J C 2013 *Phys. Rev. E* **88** 032206

Journal of Physics: Conference Series **750** (2016) 012019 doi:10.1088/1742-6596/750/1/012019

The effect of tick size on trading volume share in three competing stock markets

Shota Nagumo[1], **Takashi Shimada**[1] **and Nobuyasu Ito**[1,2]
[1] Department of Applied Physics, Graduate School of Engineering, The University of Tokyo, Hongo 7-3-1, Bunkyo-ku, Tokyo 113-8654, Japan
[2]RIKEN Advanced Institute for Computational Science, Minatojima Minamicho 7-1-26, Chuo-ku, Kobe, Hyogo 650-0047, Japan
Email: shs-nagu@serow.t.u-tokyo.ac.jp

Abstract. The relationship between tick sizes and trading volume share in two and three competing markets is studied theoretically. By introducing a simple model which is equipped with multiple markets and non-strategic traders, we analytically calculate the share. It is shown that share is shifted from a market with a larger tick size to a market with a smaller tick size, and the size of share-shift is determined by difference between tick sizes not by ratio between tick sizes in both cases of two markets and three markets.

1. Introduction

Multiple stock markets coexist in one country and compete with each other in many cases. For example, 13 stock markets exist in the U.S.[1]. Also in Japan, other than Tokyo Stock Exchange, two Proprietary Trading Systems (PTS), which are opened by online stock trading companies exist. One property different among markets is "tick size". Tick size is a unit of order price. It is expected that a smaller tick size allows the traders buy and sell at better price and hence leads to the increase in market share[2]. Actually in Japan, by the end of 2013, PTS got about 10% of share in total trading of Nikkei average adoption stocks. And then Tokyo Stock Exchange reduced tick sizes of 100 stock brands in the first section in January and July, 2014. After reducing tick sizes, Tokyo Stock Exchange recovered its market share[3].

Mizuta *et al.* conducted a simulation study on effect of tick size on trading share by using a market model which is equipped with two markets with different tick sizes and strategic traders. It was found that a market with a larger tick size is deprived of its initial major share by the competitive market, but a market with a larger tick size keeps its share at least to some extent even if a tick size of the competitive market is infinitely small[4]. Nagumo *et al.* explained the mathematical background of the above result by showing that the size of share-shift is determined by difference between tick sizes, not ratio between tick sizes, using a simple model with two markets and non-strategic traders[5].

In this paper, we will expand this simple model with two markets into a model with three markets. Using the model, we examine whether the mathematical background of share-shift in case of three markets is consistent with that of two markets.

2. Model

There are markets which have different tick sizes. We assume only selling orders. We also assume that these prices are fluctuated randomly in the markets from a "ideal potential price". This ideal potential price is assumed to be always 0, and the fluctuation of the price is assumed to be determined by standard uniform distribution $U(0, 1)$. The fraction of the prices under tick sizes are rounded up.

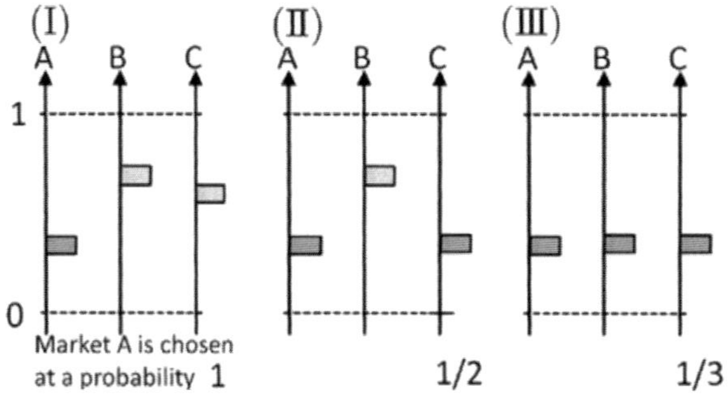

Fig. 1: Model in case of three markets. The axes represent prices. The boxes represent orders. In case of (I), market A uniquely has the cheapest price, so market A is chosen at a probability 1. In case of (I I), market A and C have the cheapest price, so market A is chosen at a probability 1/2. In case of (III), all the price are the same, so market A is chosen at a probability 1/3.

Comparing the prices of markets, share of market X is defined according with the following rule. If a price of at least one market is cheaper than the price of market X, market X is not chosen. If the price of market X is the cheapest among the markets, and n other markets also have the same price ($n \geq 0$), market X is chosen at a probability $\frac{1}{n+1}$. The probability at which market X is chosen is defined as its share.

Tick sizes of the markets are based on 1 which is the standard deviation of standard uniform distribution. In case of two markets, we define tick sizes of market A and B as a and b. We suppose that a, b and 1 satisfy $0 < b \leq a \leq 1$ and have multiple relations with the others. In the same way, in case of three markets, we define tick sizes of market A, B and C as a, b and c. We suppose that a, b, c and 1 satisfy $0 < c \leq b \leq a \leq 1$ and have multiple relations with the others.

The schematic diagrams of the model are shown in Fig. 1. This pictures show probabilities at which market A is chosen for each case of (I), (II) and (III) in case of three markets.

3. Results

Firstly, we review the result of the case of two markets[5], and next, we argue on the case of three markets.

3.1. Two markets

We define the prices in market A and B as p_A and p_B. Those size relations are divided into

$$\text{(i)}\, p_A < p_B, \quad \text{(ii)}\, p_B < p_A, \quad \text{(iii)}\, p_A = p_B.$$

The probabilities of each case are defined as P_1', P_2' and P_3'. They are derived as follows. We define p_A and p_B as $p_A = ma$ and $p_B = nb$. m is varied for $m = 1, 2, \cdots, 1/a$, and each probability is a. n is varied for $n = 1, 2, \cdots, 1/b$, and each probability is b. By these definitions, P_1' and P_3' represent a probability that $n > \frac{ma}{b}$ is satisfied and a probability that $n = \frac{ma}{b}$ is satisfied, respectively. Therefore, P_1', P_3' and P_2' are

$$P_1' = \sum_{m=1}^{1/a} a \sum_{n=1+ma/b}^{1/b} b = \frac{1-a}{2}, \tag{1}$$

$$P_3' = \sum_{m=1}^{1/a} ab = b, \tag{2}$$

$$P_2' = 1 - P_1 - P_3 = \frac{1+a-2b}{2}. \tag{3}$$

Market A is chosen at a probability 1 in case of (i) and 1/2 in case of (iii). Likewise, market B is chosen at a probability 1 in case of (ii) and 1/2 in case of (iii) Therefore, share of market A and B should be

$$\begin{aligned} S_A^* &= 1 \cdot P_1' + \frac{1}{2}P_3' \\ &= \frac{1}{2} - \frac{1}{2}(a-b), \end{aligned} \tag{4}$$

$$\begin{aligned} S_B^* &= 1 \cdot P_2' + \frac{1}{2}P_3' \\ &= \frac{1}{2} - \frac{1}{2}(b-a). \end{aligned} \tag{5}$$

Therefore, it is found that share is shifted from a market with a larger tick size to a market with a smaller tick size. Moreover, the size of share-shift is determined by difference between tick sizes, not ratio between tick sizes.

3.2. *Three markets*

We define the prices in market A, B and C as p_A, p_B and p_C. Those size relations are divided into

(i) $p_A < p_B, p_C$, (ii) $p_B < p_A, p_C$, (iii) $p_C < p_A, p_C$,
(iv) $p_B = p_C < p_A$, (v) $p_C = p_A < p_B$, (vi) $p_A = p_B < p_C$,
(vii) $p_A = p_B = p_C$.

The probabilities of each case are defined as P_1, P_2, $\cdots$, P_7, respectively. They are derived as follows. We define p_A, p_B and p_C as $p_A = la, p_B = mb$ and $p_C = nc$. l is varied for $l = 1, 2, \cdots, 1/a$, and each probability is a. Likewise, m and n are varied for $m = 1, 2, \cdots, 1/b$ and $n = 1, 2, \cdots, 1/c$, and each probability is b and c. By these definitions, for example, P_1 represents a probability that $m > \frac{la}{b}, n > \frac{la}{c}$ is satisfied. Therefore, P_1 is

$$\begin{aligned} P_1 &= \sum_{l=1}^{1/a} a \sum_{m=1+la/b}^{1/b} b \sum_{n=1+la/c}^{1/c} c \\ &= \frac{1}{6}(1-a)(2-a). \end{aligned} \tag{6}$$

In the same way, $P_2, \cdots, P_7$ are derived as

$$P_2 = \frac{1}{12}(4 + 3a - 9b - a^2 + 3ab), \tag{7}$$

$$P_3 = \frac{1}{12}(4 + 3a + 3b - 12c - a^2 + 3ab), \tag{8}$$

$$P_4 = \frac{c}{2}(1 + a - 2b), \tag{9}$$

$$P_5 = \frac{c}{2}(1 - a), \tag{10}$$

$$P_6 = \frac{b}{2}(1 - a), \tag{11}$$

$$P_7 = bc. \tag{12}$$

Market A is chosen at a probability 1 in case of (i), 1/2 in case of (v),(vi), and 1/3 in case of (vii). Likewise, market B is chosen at a probability 1 in case of (ii), 1/2 in case of (iv),(vi), and 1/3 in case of (vii), and market C is chosen at a probability 1 in case of (iii), 1/2 in case of (iv),(v), and 1/3 in case of (vii). Therefore, share of market A, B and C should be

$$S_A^* = 1 \cdot P_1 + \frac{1}{2}(P_5 + P_6) + \frac{1}{3}P_7$$
$$= \frac{1}{3} - \frac{1}{2}\left(a - \frac{b+c}{2}\right) + \frac{1}{12}(2a^2 - 3ab - 3ac + 4bc), \quad (13)$$
$$S_B^* = 1 \cdot P_2 + \frac{1}{2}(P_4 + P_6) + \frac{1}{3}P_7$$
$$= \frac{1}{3} - \frac{1}{2}\left(b - \frac{c+a}{2}\right) + \frac{1}{12}(-a^2 + 3ac - 2bc), \quad (14)$$
$$S_C^* = 1 \cdot P_3 + \frac{1}{2}(P_4 + P_5) + \frac{1}{3}P_7$$
$$= \frac{1}{3} - \frac{1}{2}\left(c - \frac{a+b}{2}\right) + \frac{1}{12}(-a^2 + 3ab - 2bc). \quad (15)$$

Using an approximation that we ignore second-order terms of a, b and c by the assumption of $a, b, c \ll 1$, S_A^*, S_B^* and S_C^* should be

$$S_A^* = \frac{1}{3} - \frac{1}{2}\left(a - \frac{b+c}{2}\right), \quad (16)$$
$$S_B^* = \frac{1}{3} - \frac{1}{2}\left(b - \frac{c+a}{2}\right), \quad (17)$$
$$S_C^* = \frac{1}{3} - \frac{1}{2}\left(c - \frac{a+b}{2}\right). \quad (18)$$

Therefore, it is found that share is shifted from a market with a larger tick size to a market with a smaller tick size. Moreover, the size of share-shift is determined by difference between tick sizes, not ratio between tick sizes. This feature is the same as the case of two markets.

4. Summary and Conclusion

We found that the size of share-shift is determined by difference between tick sizes in both cases of two markets and three markets. Therefore, in both cases of two markets and three markets, markets with any tick sizes keep at least a few amount of share even though tick sizes of the competitive markets is infinitely small.

It is known in case of two markets that even a market with a larger tick size can keep its major initial share against the competitive market with a smaller tick size if traders strongly prefer a market with large share when they choose a market[5]. It is a future task to examine whether this feature is consistent with the case of three markets if such "share-preference" introduced into the model.

References

[1] Ohta T 2014 *Daiwa Institute of Research*
[2] Uno A 2012 *Securities Analysts Journal* **50** 6
[3] Hayashi T 2015 *JPX Working Paper* **10**
[4] Mizuta T *et al* 2013 *JPX Working Paper* **2**
[5] Nagumo S *et al* 2016 *JPSJ* (*in preparation*)

Aging and related scaling during the collapse of a polymer

Suman Majumder and Wolfhard Janke
Institut für Theoretische Physik, Universität Leipzig, Postfach 100 920, 04009 Leipzig, Germany
E-mail: suman.majumder@itp.uni-leipzig.de, wolfhard.janke@itp.uni-leipzig.de

Abstract. We present results showing aging during the collapse of a polymer chain quenched from an expanded state to the globular phase via Monte Carlo dynamics of a model homopolymer. The constructed two-time correlation function, an analogue of the density-density autocorrelation, shows a power-law scaling with respect to the size of monomer clusters formed during the collapse. The numerical estimates obtained at different quench temperatures are in agreement with a theoretically predicted bound for the decay exponent governing the scaling behavior.

1. Introduction

The characteristic feature of aging is that a younger system relaxes faster than an older one. Considering its realization in diverse fields starting from spin systems to biology makes it an important as well as interesting topic of research. Aging is quite popular for systems with slow dynamics, e.g., glasses (both spin glass and structural glass) [1, 2], and especially for ordering kinetics [3] the theoretical understanding is quite developed [4, 5]. On the other hand, there have been little efforts in exploring this phenomenon in the dynamics of the collapse of a polymer chain when it is quenched from its extended state in a good solvent (above the collapse transition temperature) to the globular phase in a poor solvent (below the collapse transition temperature).

Aging in general is probed by multiple-time properties of an evolving system, e.g., the two-time autocorrelation function defined as

$$C(t,t_w) = \langle O_i(t)O_i(t_w)\rangle - \langle O_i(t)\rangle\langle O_i(t_w)\rangle, \qquad (1)$$

where t and t_w ($\ll t$) are the observation and waiting times, respectively. Here O_i is a parameter that reflects the spatio-temporal changes during the physical process, e.g., the time- and space-dependent order parameter for ordering kinetics. For ferromagnetic ordering $C(t,t_w)$ shows a power-law decay as

$$C(t,t_w) \sim (\ell/\ell_w)^{-\lambda}; \quad d/2 \leq \lambda \leq d, \qquad (2)$$

where ℓ (ℓ_w) is the growing length scale at time t (t_w) and λ is the dynamical aging exponent. The bound on λ in (2) is known as the Fisher-Huse (FH) bound, first predicted for ordering spin glasses [6] and later found to be valid for ferromagnetic ordering as well. In the latter case there exists even detailed theoretical understanding of the full form of the scaling function [7, 8, 9]. Here we exploit the similarities with ordering kinetics [10] to show the presence of aging and associated dynamical scaling during the collapse of a homopolymer.

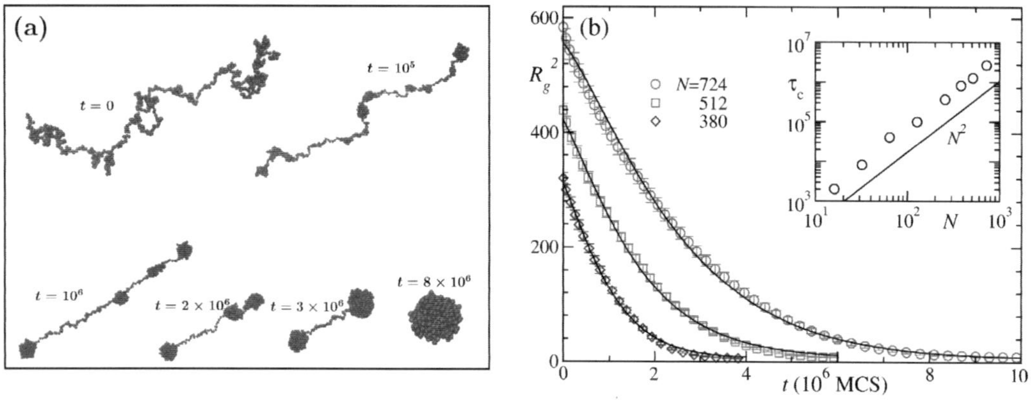

Figure 1. (a) Time evolution of a polymer with $N = 724$, after being quenched from $T_h = 10$ to $T_q = 1$. (b) Variation of the squared radius of gyration with time for three different N as indicated. The continuous lines are fits to a stretched exponential decay, described in the text. The inset shows the scaling of the collapse times τ_c extracted from the fits, as a function of the chain length. The solid line there depicts the Rouse behavior $\tau_c \sim N^2$.

2. Model and Methods

We consider the bead-spring model for a flexible homopolymer in $d = 3$ with a nonbonded monomer-monomer interaction energy $E_{\text{nb}}(r_{ij}) = E_{\text{LJ}}(\min(r_{ij}, r_c)) - E_{\text{LJ}}(r_c)$, where $E_{\text{LJ}}(r) = 4\epsilon\left[\left(\frac{\sigma}{r}\right)^{12} - \left(\frac{\sigma}{r}\right)^{6}\right]$ is the standard Lennard-Jones (LJ) potential with σ as the diameter of the monomers and ϵ being the interaction strength which is set to unity. A cut-off radius r_c $(= 2.5\sigma)$ is introduced for faster computation. The bond between successive monomers is maintained via the standard finitely extensible non-linear elastic (FENE) potential

$$E_{\text{FENE}}(r_{ii+1}) = -\frac{K}{2}R^2 \ln\left[1 - \left(\frac{r_{ii+1} - r_0}{R}\right)^2\right] \tag{3}$$

with $K = 40$, $r_0 = 0.7$, $R = 0.3$. For these choices and $\sigma = r_0/2^{1/6}$ the thermodynamics of this model is well known with no signature of glassy states [12]. Dynamics in this model is incorporated via Monte Carlo (MC) simulations where displacement moves, chosen randomly within $[-\sigma/10 : \sigma/10]$, are attempted on randomly chosen monomers following the standard Metropolis algorithm [11]. The unit of time is one MC step (MCS) that consists of N such attempted moves where N is the number of monomers present in the polymer chain. The unit of temperature is ϵ/k_B with the Boltzmann constant k_B set to unity. We have prepared the initial configurations at high temperature ($T_h = 10$) which were then quenched into the globular phase at T_q, well below the collapse transition temperature T_θ ($> T_q$). All presented results are averaged over several hundred initial realizations.

3. Results

In Fig. 1(a), we show evolution snapshots at different times during the collapse for a polymer with $N = 724$. Initially at $t = 0$ the polymer is in an expanded state with fluctuations in the density of monomers along the chain. The fluctuations eventually lead to the formation of stable nucleation clusters, evident from the configuration at $t = 10^5$. These clusters then start growing by withdrawing monomers from the bridges connecting them and once the tension in the bridge reaches maximum the clusters coalesce with each other to form bigger clusters as shown in the snapshots at intermediate times. These events continue until they give rise to a single cluster like

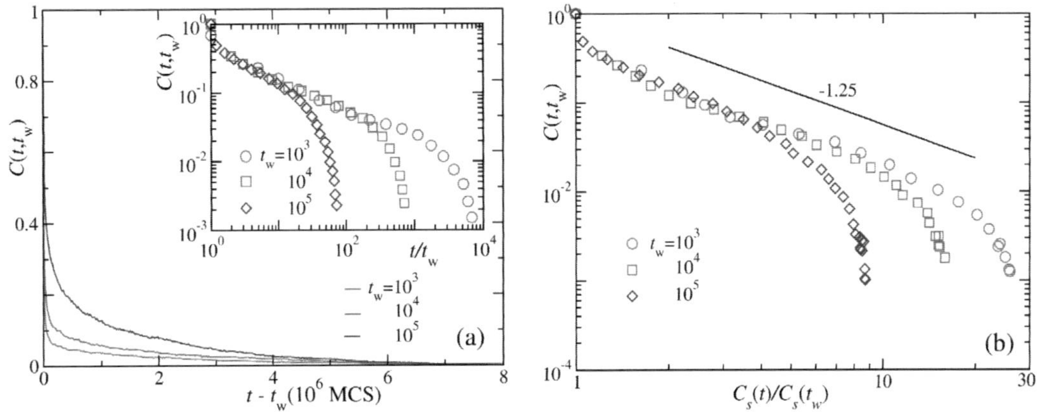

Figure 2. (a) Plot of the autocorrelation function $C(t, t_w)$ of a polymer with $N = 724$ for three different values of t_w at $T_q = 1$. The inset shows the scaling of $C(t, t_w)$ as function of t/t_w. (b) Demonstration of the dynamic scaling of $C(t, t_w)$ as a function of $C_s(t)/C_s(t_w)$. The solid line corresponds to the power-law decay (4) with an aging exponent $\lambda_c = 1.25$.

in the snapshot at $t = 8 \times 10^6$. The sequence of events observed here looks pretty similar to the phenomenological picture of Halperin and Goldbart [13]. At this point it is worth mentioning that this cluster-growth phase of the collapse bears similarities with usual coarsening systems. To define the appropriate length scale for the cluster growth, we conveniently opt to measure the average cluster size $C_s(t)$ [$\sim \ell(t)^d$]. For that we identify a cluster on the basis of the local density around a monomer i, i.e., by counting the number of monomers $n_i = \sum_{j=1}^{N} \Theta(r_c - r_{ij})$, where Θ is the Heaviside step function. We call it a cluster when $n_i \geq n_{\min}$, where $n_{\min}$ is a threshold value. It has been shown that the method is insensitive to reasonable choices of $n_{\min}$ [10]. Once the total number of discrete clusters is identified we measure the average cluster size, $C_s(t)$, as the average number of monomers present in each cluster. The kinetics of this cluster growth during the collapse has been explored by us elsewhere and was found to show a linear growth with time. As we are here particularly focused on the two-time properties, we shall not discuss the cluster growth further but refer to Ref. [10] for details.

In Fig. 1(b), we show the time evolution of the mean squared radius of gyration, $R_g^2 = \sum_{i,j}(r_i - r_j)^2/2N^2$, for different N. The solid lines there are fits to the stretched exponential function $R_g^2 = a_0 + a_1 \exp[-(t/\tau_c)^\beta]$, where $\beta \approx 1.20 - 1.25$ and τ_c gives a measure of the collapse time. In the inset of Fig. 1(b), we show the scaling of τ_c with the chain length. The data seem to be quite parallel to the solid line which represents the Rouse scaling behavior ($\sim N^2$), expected for the diffusive dynamics of polymers in highly viscous solvent neglecting hydrodynamics [14].

Moving on to the prime quantity of interest in this work, we use our cluster identification technique to define the parameter O_i, in order to calculate the two-time property according to Eq. (1). We assign $O_i = \pm 1$ depending on whether the monomer is inside (+) or outside (−) a cluster [15]. The autocorrelation function $C(t, t_w)$ calculated this way is an analogue of the density-density autocorrelation in particle systems or the order-parameter autocorrelation for spin systems. In Fig. 2(a), the main frame shows plots of such autocorrelations $C(t, t_w)$ as function of $t - t_w$, clearly indicating the absence of time-translational invariance, the evidence for aging. In the inset we show the scaling of $C(t, t_w)$ with respect to t/t_w, a signature of simple aging. Note that at late time the abrupt decay is due to the onset of finite-size effects. The

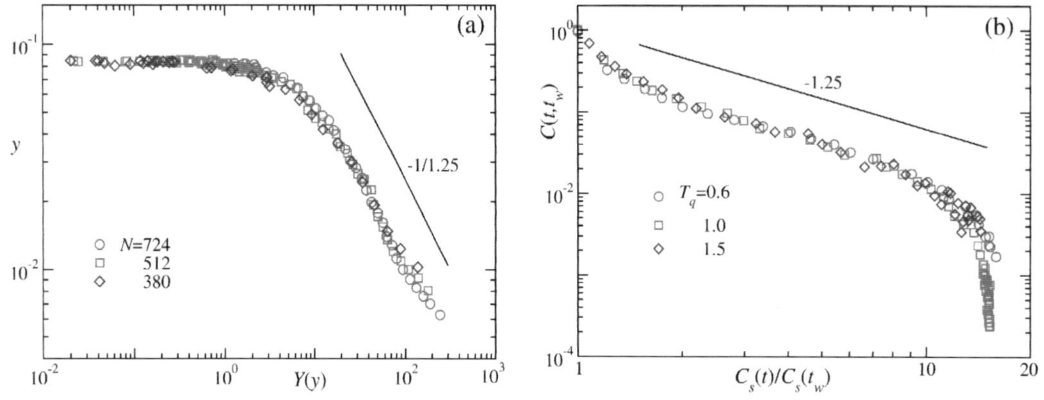

Figure 3. (a) Finite-size scaling plots for $C(t, t_w)$ with $t_w = 10^4$ at $T_q = 1$. The solid line shows the expected behavior $y \sim Y(y)^{-1/\lambda_c}$ with $\lambda_c = 1.25$. (b) Scaling plot of $C(t, t_w)$ as a function of $C_s(t)/C_s(t_w)$ for different quench temperatures T_q as mentioned.

observed scaling of the autocorrelations with respect to the growing length scale in ordering kinetics motivated us to look for such scaling also here, of course with respect to the growing cluster size $C_s(t)$, the corresponding relevant quantity. Figure 2(b) shows such a scaling plot of $C(t, t_w)$ as a function of $C_s(t)/C_s(t_w)$. The steep early-time decay is attributed to short-time fluctuations. The linear behavior of the data on a double-log scale suggests that the scaling is power law in nature which can be quantified as

$$C(t, t_w) = A\,[C_s(t)/C_s(t_w)]^{-\lambda_c}, \tag{4}$$

where λ_c is the corresponding aging exponent and A is an amplitude that depends on temperature. Below we derive a theoretical bound on λ_c.

From the fact that $C(t, t_w)$ is an analogue to the density-density autocorrelation it is justified to assume $C(t, t_w) \sim \langle \rho(t)\rho(t_w)\rangle$ where ρ is the average density of monomers. Considering t_w to be an early enough time, one can argue that the polymer is in an extended configuration so that the average density $\rho(t_w) \sim C_s/{C_s}^{d\nu}$ for a set of C_s monomers, where ν is the Flory exponent for the size, i.e., radius of gyration ($R_g \sim C_s^{\nu}$) of a polymer chain of C_s monomers. Now in the first limiting situation, at later time t, $\rho(t) = 1$, assuming all the monomers to be inside a single cluster and a maximum overlap between $\rho(t)$ and $\rho(t_w)$, provides the lower bound on λ_c with $C(t, t_w) \sim C_s/C_s^{\nu d} \sim C_s^{-(\nu d-1)}$. The other limiting situation of the polymer being still in an expanded state at time t, provides the upper bound on λ_c with $C(t, t_w) \sim C_s^{-2(\nu d-1)}$. Combining, we get the bound on λ_c as [15]

$$(\nu d - 1) \leq \lambda_c \leq 2(\nu d - 1). \tag{5}$$

Putting $\nu \approx \nu_F = 3/5$ in Eq. (5), where ν_F is the Flory approximation for polymers with excluded volume interaction, one would get $4/5 \leq \lambda_c \leq 8/5$. The consistency of our data in Fig. 2(b) with the line having slope -1.25 shows that this bound is likely to be valid. A more precise estimate of λ_c based on a finite-size scaling analysis can be done by writing down $NC(t, t_w) = Y(y)$, where $Y(y)$ is a finite-size scaling function and y is the scaling variable. Now, for an appropriate choice of $y = [C_s(t)/C_s(t_w)]\,(NA)^{-1/\lambda_c}$, one gets $Y(y) \sim y^{-\lambda_c}$, i.e., $y \sim Y(y)^{-1/\lambda_c}$ in the scaling regime. By fixing the value of $A = 1$ we have tuned the value of λ_c to obtain the optimum collapse of data from different N. In this exercise a reasonably good data collapse is obtained for $\lambda_c = 1.25(5)$, consistent with the predicted bound (5). A representative of such finite-size scaling plots with $\lambda_c = 1.25$ is shown in Fig. 3(a). The data from different system sizes collapse onto each other to follow the scaling behavior $y \sim Y^{-1/\lambda_c}$ [15]. The plateau for smaller values of Y is due to the onset of finite-size effects at late times.

In Fig. 3(b) we demonstrate the independence of the autocorrelation function on the quench temperature T_q, where the $y-$axis has been multiplied by an appropriate factor in order to take care of the amplitudes A at different T_q. The collapse of the data from different T_q onto a single master curve shows that the power-law scaling is independent of the quench depth. Note that in this plot we have used different t_w for different T_q such that the values of $C_s(t_w)$ are approximately the same and consequently the data for all T_q have the same scaling width. The solid line shows the consistency of our data with the value of the aging exponent obtained from the finite-size scaling exercise at $T_q = 1$ and implies that the bound (5) is in fact valid at all temperatures.

4. Conclusion

We have presented results from the diffusive dynamics of a homopolymer collapse via Monte Carlo simulation seeking evidences for aging. By constructing a suitable two-time correlation function we have captured the aging during the collapse and also found an associated power-law scaling with respect to the growing clusters of monomers. In addition, for the decay exponent governing the scaling we have derived a theoretical bound which is not only found to be valid when compared with our estimates using finite-size scaling analysis but also appears to be independent of the quench temperatures.

Acknowledgments

The work was funded by the DFG (German Science Foundation) under Grant No. JA 483/24-3 and SFB/TRR 102 (project B04), and further supported by the EU through the Marie Curie IRSES network DIONICOS under Contract No. PIRSES-GA-2013-612707.

References

[1] Bouchaud J-P, Cugliandolo L F, Kurchan J and Mézard M 1997 *Spin Glasses and Random Fields* ed A P Young (Singapore: World Scientific) p 161
[2] Kob W 2000 *Soft and Fragile Matter* ed M E Cates and M Evans (Bristol: IOP) p 259
[3] Bray A J 2002 *Adv. Phys.* **51** 481
[4] Zannetti M 2009 *Kinetics of Phase Transitions* ed S Puri and V Wadhawan (Boca Raton, FL: CRC Press) p 153
[5] Henkel M and Pleimling M 2010 *Non-Equilibrium Phase Transitions, Vol 2: Ageing and Dynamical Scaling Far from Equilibrium* (Heidelberg: Springer)
[6] Fisher D S and Huse D A 1988 *Phys. Rev.* B **38** 373
[7] Henkel M, Picone A and Pleimling M 2004 *Europhys. Lett.* **68** 191
[8] Lorenz E and Janke W 2007 *Europhys. Lett.* **77** 10003
[9] Midya J, Majumder S and Das S K 2014 *J. Phys.: Condens. Matter* **26** 452202
[10] Majumder S and Janke W 2015 *Europhys. Lett.* **110** 58001
[11] Landau D P and Binder K 2005 *A Guide to Monte Carlo Simulations in Statistical Physics* (Cambridge: Cambridge University Press)
[12] Schnabel S, Bachmann M and Janke W 2009 *J. Chem. Phys.* **131** 124904
[13] Halperin A and Goldbart A 2000 *Phys. Rev.* E **61** 565
[14] Rouse E 1953 *J. Chem. Phys.* **21** 1272
[15] Majumder S and Janke W 2016 *Phys. Rev.* E **93** 032506

Journal of Physics: Conference Series **750** (2016) 012021 doi:10.1088/1742-6596/750/1/012021

Quantum Decoherence and Thermalization at Finite Temperatures of Non-Degenerate Spin Systems via Small Spin Environments

M.A. Novotny
Dept. of Physics and Astronomy and HPC^2 Center for Computational Sciences, Mississippi State U., Mississippi State, MS 39762, USA

F. Jin
Institute for Advanced Simulation, Jülich Supercomputing Centre, Forschungszentrum Jülich, D-52425 Jülich, Germany

H. De Raedt
Zernike Institute for Advanced Materials, University of Groningen, Nijenborgh 4, NL-9747AG Groningen, The Netherlands

K. Michielsen
Institute for Advanced Simulation, Jülich Supercomputing Centre, Forschungszentrum Jülich, D-52425 Jülich, Germany
RWTH Aachen University, D-52056 Aachen, Germany

E-mail: man40@msstate.edu

Abstract. We study the case of a small quantum spin system S with a non-degenerate groundstate coupled to a small quantum spin bath. Finite temperature measures for both quantum decoherence and thermalization are studied. The computational results, obtained from exact diagonalization, compare well with a recent perturbation theory prediction, even when the system and bath are of comparable sizes.

1. Introduction

In quantum statistical mechanics a quantum system S is coupled to a quantum environment (or bath) E, and the time evolution of the entirety $S+E$ is governed by the time-dependent Schrödinger equation. A recent perturbation theory prediction [1] assumes that initially the entirety is in a canonical-thermal state [2, 3, 4], and makes predictions for both a measure σ for the decoherence and a measure δ for the thermalization. In this short paper the case of a non-degenerate ground state for S is examined for spin systems coupled to small spin baths. This study is performed in order to test the theoretical predictions for σ and δ for environments comparable in size to the system.

2. Theory and Results

The Hamiltonian of the entirety $S+E$ is taken to be

$$\mathcal{H} = \mathcal{H}_S + \mathcal{H}_E + \lambda \mathcal{H}_{SE} \,, \tag{1}$$

where $\mathcal{H}_S$ and $\mathcal{H}_E$ are the system and environment Hamiltonian, respectively and $\mathcal{H}_{SE}$ describes the interaction between the system S and the environment E. We consider only time-independent Hamiltonians. The global system-environment coupling strength is λ.

In the normal fashion [5, 6, 7, 8], the quantum system S is described by the reduced density matrix $\widetilde{\rho}(t) \equiv \mathbf{Tr}_E \rho(t)$, where $\rho(t) = |\Psi(t)\rangle \langle \Psi(t)|$ is the density matrix of the entirety $S+E$ at time t, $\Psi(t)$ denotes the wavefunction of the entirety, and $\mathbf{Tr}_E$ denotes the trace over all degrees of freedom of the environment. We assume that $\widetilde{\rho}$ is expressed in the basis in which $\mathcal{H}_S$ is diagonal.

The degree of decoherence of the system is characterized by [9, 10]

$$\sigma(t) = \sqrt{\sum_{i=1}^{D_S-1} \sum_{j=i+1}^{D_S} |\widetilde{\rho}_{ij}(t)|^2} \,, \tag{2}$$

where $\widetilde{\rho}_{ij}(t)$ is the matrix element (i,j) of the reduced density matrix in the basis that diagonalizes H_S, and D_S is the dimension of the Hilbert space of $\mathcal{H}_S$. We also define a quantity measuring thermalization, namely the difference between $\widetilde{\rho}_{ii}$ and the canonical distribution as [9]

$$\delta(t) = \sqrt{\sum_{i=1}^{D_S} \left(\widetilde{\rho}_{ii}(t) - e^{-b(t)E_i^{(S)}} \Big/ \sum_{i'=1}^{D_S} e^{-b(t)E_{i'}^{(S)}} \right)^2} \,, \tag{3}$$

where $\{E_i^{(S)}\}$ denote the eigenvalues of $\mathcal{H}_S$ and $b(t)$ is a fitting parameter. For excellent fits to the canonical ensemble, the fitting parameter $b(t)$ should approach the inverse temperature $\beta = 1/T$ (in units $k_B = 1$) at large times.

Assuming that initially the entirety is in a canonical-thermal state, Ref. [1] gives a perturbation theory result valid for small $\beta\lambda$, for the expectation value for the decoherence measure of Eq. (2)

$$\begin{aligned} \mathcal{E}\left(\sigma^2\right) &= \frac{1}{2}\frac{Z_E(2\beta)}{Z_E^2(\beta)}\left(1 - \frac{Z_S(2\beta)}{Z_S^2(\beta)}\right) - \frac{2\,D}{(D+1)}\frac{Z_E(3\beta)}{Z_E^3(\beta)}\left(\frac{Z_S(2\beta)}{Z_S^2(\beta)} - \frac{Z_S(3\beta)}{Z_S^3(\beta)}\right) \\ &\quad + \frac{3}{2}\frac{D}{(D+1)}\frac{Z_E^2(2\beta)}{Z_E^4(\beta)}\left(\frac{Z_S(2\beta)}{Z_S^2(\beta)} - \frac{Z_S^2(2\beta)}{Z_S^4(\beta)}\right) . \end{aligned} \tag{4}$$

Equation (4) is written in terms of partition functions for S and E at inverse temperature β, respectively $Z_S(\beta)$ and $Z_E(\beta)$. Here $D = D_S D_E = 2^{N_S+N_E}$ is the dimension of the Hilbert space of the entirety, with N_S (N_E) the number of spin-1/2 particles in the system (environment with Hilbert space dimension D_E). It is important to remember that even when $\lambda=0$ (no part of the Hamiltonian coupling the system to the environment) the assumption that the entirety is in a canonical-thermal state [1] means that the entirety wavefunction is not a direct product of wavefuctions from S and E. The groundstate degeneracy of the system (environment) is g_S (g_E). In this paper we concentrate on non-degenerate groundstates for S, so $g_S=1$. We restrict ourselves to $g_S=1$ since in the limit of low temperatures [1] one has

$$\mathcal{E}\left(\sigma^2\right) = \frac{1}{2}\frac{g_S - 1}{g_E g_S}\left(1 - \frac{D}{(D+1)\,g_E g_S}\right) \tag{5}$$

and thus for low temperatures $\mathcal{E}(\sigma^2) \to 0$ for g_S=1.

We performed exact diagonalization on an entirety $S+E$ composed of a ring of spin-1/2 particles (spins). The exact diagonalization allows us to calculate exactly the partition functions Z_S and Z_E of Eq. (4). Therefore direct comparisons between the theoretical results and the numerical results could be performed with **no** adjustable parameters. All couplings within the ring are made randomly between $[-1, 1]$, the explicit random values are listed in Table 1, for nearest-neighbor spin couplings in a ring of N=N_S+N_E spins with periodic boundary conditions. The Hamiltonian between spins j and j+1 is $\mathcal{H}_{j,j+1} = -J^x_{j,j+1} s^x_j s^x_{j+1} - J^y_{j,j+1} s^y_j s^y_{j+1} - J^z_{j,j+1} s^z_j s^z_{j+1}$.

In this paper we keep N_S=4, and always keep the interactions in $\mathcal{H}_S$ as the couplings in Table 1 for $\mathcal{H}_{j,j+1}$ with $j = 1, 2, 3$. Specifically, $\mathcal{H}_S = \mathcal{H}_{1,2} + \mathcal{H}_{2,3} + \mathcal{H}_{3,4}$, and the degeneracy of the ground state is thus g_S=1 always. The system-environment Hamiltonian used in this paper is $\mathcal{H}_{SE} = \mathcal{H}_{4,5} + \mathcal{H}_{N,1}$, always with the values listed in Table 1 (note the periodic boundary conditions since we study a ring of spins).

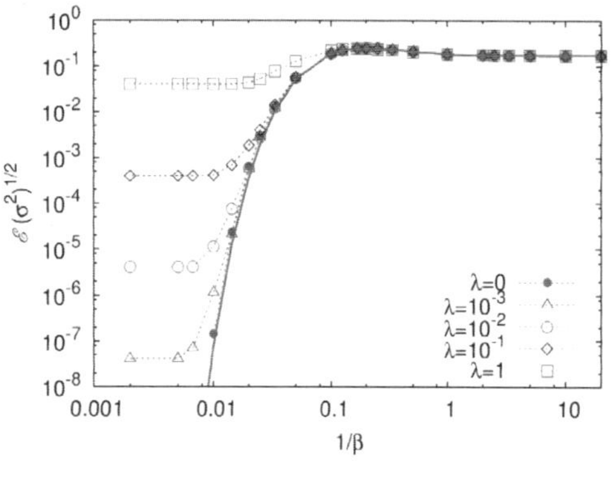

Figure 1. The square root of the expectation value for the decoherence measure σ^2 for a spin system with N_S=4 and g_S=1 coupled to a spin bath with N_E=4 and g_E=1 as a function of the temperature $1/\beta$ for various values of the global system-environment coupling strength λ. The solid line is obtained from Eq. (4). The dashed lines are guides to the eye.

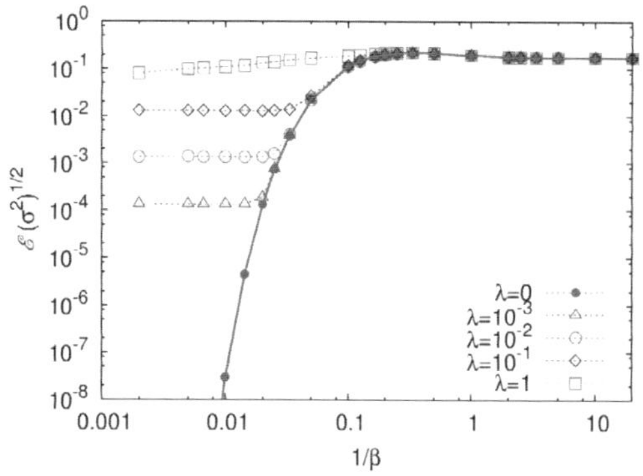

Figure 2. The same as in Fig. 1. Here, however, all interaction strengths in the $N_E = 4$ spin Hamiltonian $\mathcal{H}_E$ are set to -1, explicitly $\mathcal{H}_E = -\sum_{j=5}^{7}\left(s^x_j s^x_{j+1} + s^y_j s^y_{j+1} + s^z_j s^z_{j+1}\right)$, and therefore the environment degeneracy is $g_E = 5$. The solid line is the perturbation theory result from Eq. (4), while the dashed lines are guides to the eye.

Figure 1 shows results for the square root of the expectation value $\mathcal{E}(\sigma^2)$ as a function of temperature, $1/\beta$, for different strengths of the system-environment coupling strength λ. The groundstate degeneracies are g_S=g_E= 1. Here there are four spins in both the system and the environment. The environment Hamiltonian here is $\mathcal{H}_E = \sum_{j=5}^{7} \mathcal{H}_{j,j+1}$ with the interaction strengths listed in Table (1). Figure 1 shows an extremely good agreement with the theory of Eq. (4), even though the dimensions of the Hilbert space of the environment and the system are equal D_E=D_S= 2^4. This is true for high temperatures, and continues to be true as long as $\beta\lambda \ll 1$. Therefore, at all temperatures, it is not required to have a very large spin bath in order to obtain the predicted value for the decoherence measure σ.

The data points in all the figures are obtained by first picking uniformly at random a typical unit vector from the entirety Hilbert space. In particular, a unit vector $|\Psi_0\rangle = \sum_{j=1}^{D_S}\sum_{p=1}^{D_E} d_{j,p}\,|j,p\rangle$ is picked with the coefficients $\{d_{j,p}\}$ complex Gaussian random numbers

Table 1. The Hamiltonian $\mathcal{H}_{j,j+1}$ interaction strengths used in this paper.

j	$j+1$	$J^x_{j,j+1}$	$J^y_{j,j+1}$	$J^z_{j,j+1}$
1	2	0.99464	0.18950	-0.83949
2	3	0.66019	-0.74690	0.46002
3	4	0.87474	0.10580	-0.93014
4	5	0.76983	0.86751	0.29104
N	1	-0.35312	0.61078	0.17238
5	6	0.16682	-0.65712	-0.69131
6	7	-0.84125	-0.36024	0.97012
7	8	-0.30384	0.29020	0.77970
8	9	-0.99592	-0.33191	0.37434
9	10	0.49823	0.25744	-0.76540
10	11	0.64452	-0.99678	-0.78684
11	12	-0.65810	-0.89997	-0.41505

normalized so that $\sum_{j=1}^{D_S}\sum_{p=1}^{D_E}|d_{j,p}|^2 = 1$. The vectors $|j,p\rangle$ are any orthogonal basis that spans the D dimensional Hilbert space of the entirety. The associated canonical-thermal state, drawn uniformly from the ensemble of all canonical-thermal states at inverse temperature β, is given by

$$|\Psi_\beta\rangle = e^{-\beta\mathcal{H}/2}\,|\Psi_0\rangle \Big/ \langle\Psi_0|\, e^{-\beta\mathcal{H}}\,|\Psi_0\rangle^{\frac{1}{2}} \;. \tag{6}$$

This is called a canonical-thermal state because for any operator for an observable $\mathcal{A}$ one has

$$\langle\Psi_\beta|\,\mathcal{A}\,|\Psi_\beta\rangle \approx \mathrm{Tr}\left(\mathcal{A}e^{-\beta\mathcal{H}}\right)\Big/\mathrm{Tr}\left(e^{-\beta\mathcal{H}}\right) \tag{7}$$

with the error in the approximation being of the order of $D^{-\frac{1}{2}}$. Each point in all the figures is an average over 10^3 vectors, as in Eq. (6), from the canonical-thermal-state ensemble.

Figure 2 shows results for the square root of the expectation value $\mathcal{E}(\sigma^2)$ as a function of temperature, $1/\beta$, for different strengths λ. Here the Hamiltonians $\mathcal{H}_S$ and $\mathcal{H}_{SE}$ are the same as in Fig. 1, but the environment has uniform, isotropic bond strengths

$$\mathcal{H}_E \;=\; -\sum_{j=N_S+1}^{N-1}\left(s_j^x s_{j+1}^x + s_j^y s_{j+1}^y + s_j^z s_{j+1}^z\right) \tag{8}$$

with N_E=4 and N=8. Consequently g_E=5. Again we observe that even though N_S=N_E=4 there is excellent agreement between the theory and the calculations as long as $\beta\lambda \ll 1$.

Figure 3 shows results for the square root of the expectation value $\mathcal{E}(\sigma^2)$ as a function of temperature, $1/\beta$, for different strengths λ for the case N_E=8, again with random couplings drawn between $[-1,1]$ for the environment couplings. Again $\mathcal{H}_S$ and $\mathcal{H}_{SE}$ are unchanged from Fig. 1, but the environment Hamiltonian is now $\mathcal{H}_E = \sum_{j=5}^{11}\mathcal{H}_{j,j+1}$ with the interaction strengths listed in Table (1). Again we observe that even though N_S=$N_E/2$ there is excellent agreement between the theory and the calculations as long as $\beta\lambda \ll 1$. In fact, there is very little difference that can be seen by eye in comparing Figure 1 and Figure 3. Note, however, that for high temperatures ($\beta \to 0$) one has $\mathcal{E}\left(\sigma^2\right) \to \left(D_S - 1\right)/2\left(D_S D_E + 1\right)$ [10], which explains the slightly different infinite temperature asymptotic values between the two figures.

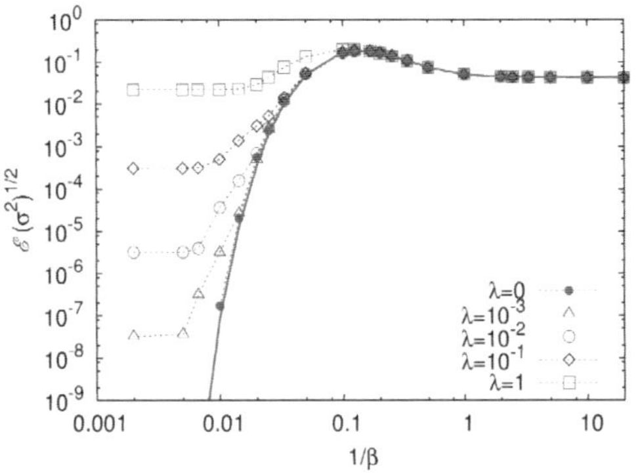

Figure 3. The same as in Fig. 1. Here, however, N_E=8 and again the environment degeneracy is g_E=1. The solid line is the perturbation theory result from Eq. (4), while the dashed lines are guides to the eye.

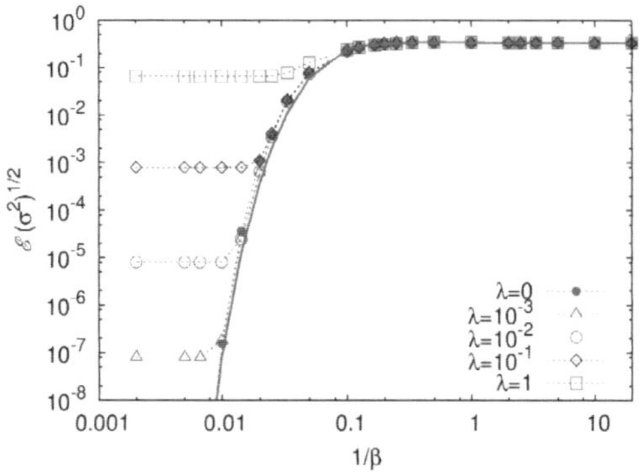

Figure 4. The same as in Fig. 1. Here, however, N_E=2 and again the environment degeneracy is g_E=1. The solid line is the perturbation theory result from Eq. (4), while the dashed lines are guides to the eye.

Figure 4 shows results for the square root of the expectation value $\mathcal{E}(\sigma^2)$ as a function of temperature, $1/\beta$, for different strengths λ for the case N_E=2. Again $\mathcal{H}_S$ and $\mathcal{H}_{SE}$ are unchanged from Fig. 1, but the environment Hamiltonian is now $\mathcal{H}_E = \mathcal{H}_{5,6}$ with the interaction strength listed in Table (1). Again we observe that even though N_S=$2N_E$ and N_E=2, there is still agreement between the theory and the calculations as long as $\beta\lambda \ll 1$. Now, however, there is a larger difference at temperatures below $1/\beta \approx 0.1$ between the perturbation theory and the results from Eq. (4). This suggests that higher order terms in the perturbation theory are relevant for small spin systems with $N_S > N_E$.

There is a related prediction for the expectation value of the measure of thermalization to that of Eq. (4) [1]. Figures 5 and 6 shows results for the square root of the expectation value of δ^2 for the same systems as studied in Fig. 3 and Fig. 4, respectively. We observe very nice agreement between the theory and the calculations as long as the temperature is higher than about $1/\beta \approx 0.1$.

3. Discussion and Conclusions

We have studied decoherence and thermalization of quantum systems, focusing on the case where the groundstate of the system is non-degenerate and the bath is comparable to the system in terms of the sizes of the Hilbert spaces. This study, together with that of Ref. [1], may be of interest to the recently developing technology of adiabatic quantum computing [11, 12], as exemplified for example by D-Wave machines [12, 13, 14].

We have compared the theoretical predications of Ref. [1] with exact diagonalization calculations. We have concentrated on spin chains, with N_S=4 spins in the system and with N_E=2, 4, and 8 spins in the environment. In all cases at high temperatures excellent agreement is found between the exact diagonalization calculations and the perturbation theory predications of Ref. [1]. The same is true at low temperatures, excellent agreement is found, provided that $\beta\lambda \ll 1$. At low temperatures, however, the N_E=2 case shows qualitative differences with the theoretical predictions. It is to be emphasized that there are no adjustable parameters in our comparison between the exact diagonalization calculations and the theoretical predictions.

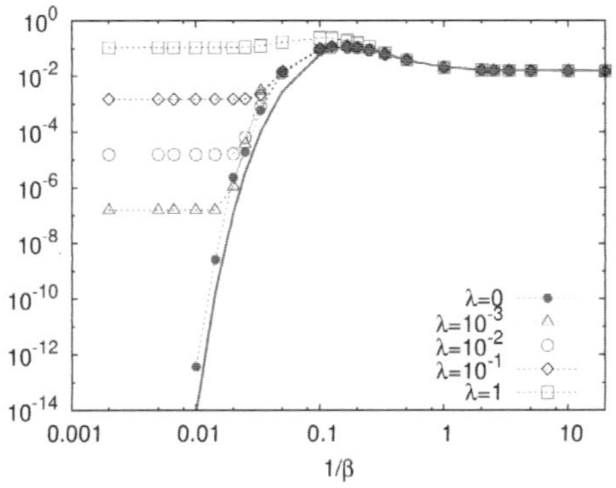

Figure 5. The square root of the expectation value for the thermalization measure δ^2 for a spin system with N_S=4 and g_S=1 coupled to a spin bath with N_E=8 and g_E=1. The entirety is the same as in Fig. 3. The solid line is the perturbation theory result, while the dashed lines are guides to the eye.

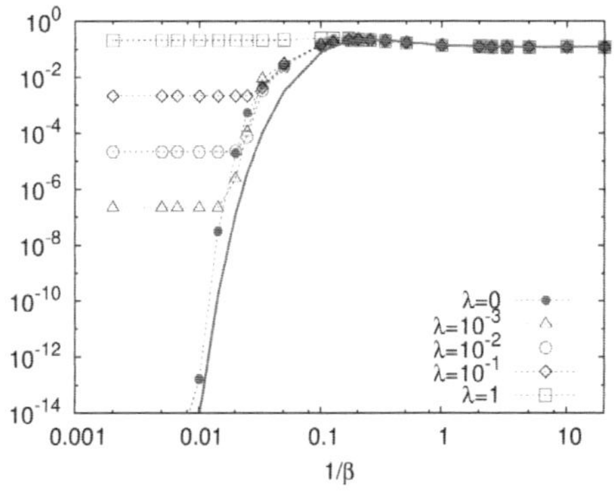

Figure 6. The square root of the expectation value for the thermalization measure δ^2 for a spin system with N_S=4 and g_S=1 coupled to a spin bath with N_E=2 and g_E=1. The entirety is the same as in Fig. 4. The solid line is the perturbation theory result, while the dashed lines are guides to the eye.

Acknowledgments

MAN is supported in part by Pacific Northwest National Laboratories.

[1] Novotny MA, Jin F, Yuan S, Miyashita S, De Raedt H, and Michielsen K, 2016, *Phys. Rev. A* **93**, 032110.
[2] Hams A. and De Raedt H, 2000, *Phys. Rev. E* **62**, 4365.
[3] Jin F, De Raedt H, Yuan S, Katsnelson M.I, Miyashita S., and Michielsen K, 2010, *J. Phys. Soc. Jpn.* **79**, 124005.
[4] Sugiura S. and Shimizu A, 2013, *Phys. Rev. Lett.* **111**, 010401.
[5] Kubo R, Toda M, and Hashitsume N, *Statistical physics II: Nonequilibrium statistical mechanics*, (Springer-Verlag, 1985).
[6] Ponomarev A.V, Denisov S, and Hänggi P, 2011, *Phys. Rev. Lett.* **106**, 010405.
[7] Ponomarev A.V, Denisov S, Hänggi P, and Gemmer J, 2012, *Europhys. Lett.* **98**, 40011.
[8] Swendsen R.H, *An Introduction to Statistical Mechanics and Thermodynamics*, (Oxford University Press, 2012).
[9] Yuan S, Katsnelson M.I, and De Raedt H, 2009, *J. Phys. Soc. Jpn.* **78** 094003.
[10] Jin F, Michielsen K, Novotny M. A, Miyashita S, Yuan S, and De Raedt H, 2013, *Phys. Rev. A* **87**, 022117.
[11] Rieffel E and Polak W, *Quantum Computing: A Gentle Introduction*, (MIT Press, 2011).
[12] Boixo S, Rønnow T.F, Isakov S.V, Wang Z, Wecker D, Lidar D.A, Martinis J.M, and Troyer M, 2014, *Nature Physics* **10**, 218.
[13] Novotny M.A, Hobl Q.L, Hall J.S, and Michielsen K, 2016, *J. Physics: Conf. Series* **681**, 012005.
[14] D-Wave, The Quantum Computing Company, http://www.dwavesys.com

Journal of Physics: Conference Series **750** (2016) 012022 doi:10.1088/1742-6596/750/1/012022

Orbital behavior around a nonuniform celestial body

Z Rosson, F Hall and T Vogel

Department of Physics, Stetson University, DeLand, FL 32723, USA

Email: zachrosson@gmail.com

Abstract. To effectively model the orbit around a nonuniform celestial body, detailed information needs to be determined of the perturbations. This research looked at one of the most crucial perturbations: the nonuniform gravitational field of a celestial body. Given an orbiting particle (a satellite), we utilized numerical methods to calculate its orbit in two dimensions around a discretized center mass structure. The gravitational acceleration imposed on the particle due to each mass point sums vectorally as the particle completes each infinitesimal time step of one orbit. There are noticeable effects on the orbit as the conditions of the center mass change. The development of a simulation code allows for the modelling of the orbit about an irregular body with satisfactory accuracy.

1. Introduction

The gravitational effects due to nonuniformities have been crucial to understanding orbits around the Moon, and the presence of mass-concentrations ("mascons") has been the focus of NASA's recent GRAIL (Gravity Recovery and Interior Laboratory) missions [1]. Over billions of years, mascons formed on the surface of the Moon as lunar material melted due to the heat of asteroids' impact, which changed the density of the surface rock. This caused an increase in the gravitational field above the mascons due to the increased mass density. Additionally, the European Space Agency's recent Rosetta mission, (the successful orbit around the 67P/Churyumov-Gerasimenko comet), served as a source of motivation for this research due to the comet's irregular shape and its particularly nonuniform gravitational field.

Orbital course adjustments must be made to account for these gravitational perturbations due to shape or mass density irregularities. In this research, we model these anomalies and simulate orbits around the nonuniform celestial bodies. For this, advanced numerical methods and gravitational potential functions have been used [2], but we attempt to approach the physical situation in a more straightforward manner while still keeping sufficient computational accuracy. This is not only because valuable orbital information due to nonuniformities can still be garnered using a more simple discretization method, but because this approach presents a novel teaching tool for students to understand and visualize orbits around more complex celestial bodies.

2. Methods

2.1 Basic physics and algorithm

The fundamental law guiding the physics of our research was Isaac Newton's Universal Law of Gravitation (1687),

$$F = -\frac{GMm}{r^2}$$

where G is the gravitational constant, M is the mass of the center body, m is the mass of the orbiting satellite, and r is the distance separating the two bodies. The inverse square distance dependence gives rise to the elliptical behavior which one expects in gravitational orbits. For the purposes of modeling, we utilized a two-dimensional plane, since relevant orbital motion in three dimensions can often be reduced to motion in a suitably chosen orbital plane. We determined the x and y components of the acceleration experienced by the orbiting particle (making the assumption that its mass is negligible compared to the center body):

$$a_x = -\frac{GMx}{(x^2+y^2)^{3/2}}$$
$$a_y = -\frac{GMy}{(x^2+y^2)^{3/2}}$$

Then, to represent the center body, we utilized a flood-fill algorithm which populates a grid of mass points based on a defined shape and point density. To account for all these infinitesimal mass contributions, we summed vectorally the acceleration due to each of the n center mass points in the discretized grid:

$$a_{nx,\,tot} = \sum_{i=1}^{n} a_{ix}$$
$$a_{ny,\,tot} = \sum_{i=1}^{n} a_{iy}$$

The orbiting particle was given an initial y-velocity and x-position at the start of its orbit, and we utilized the structure of the Euler-Cromer algorithm to solve these differential equations for its equations of motion [3]:

$$v_x = v_x' + a_{nx,tot} * dt$$
$$v_y = v_y' + a_{ny,tot} * dt$$
$$x = x' + v_x * dt$$
$$y = y' + v_y * dt$$

The prime notation represents the velocity and position values from the previous time step to be used in the current computation. Essentially, the cumulative effect of the gravitational acceleration from each mass point is used to actively compute the velocity and position of the orbiting particle for every infinitesimal time step (*dt*). This process is performed iteratively until one full orbital period has passed. Our procedure deviates from the Euler-Cromer source equations only with the aforementioned summation of acceleration. Furthermore, we plotted the orbit using lists of x and y coordinates compiled during the iterations and printed values of acceleration, velocity, position, and time within small intervals at apoapsis and periapsis to test basic orbital energy conservation (i.e. convergence of initial conditions).

2.2 Justifications

To achieve sufficient accuracy with this method, the infinitesimal time step had to be chosen with regard to its ability to close an orbit but also to limit excessive computational run times. Table 1 shows a basic error analysis of a simple physical system to demonstrate this algorithm's accuracy for one orbital period at increasingly small time steps.

	Time (days)	x-position (km)	y-position (km)	x-velocity (km/day)	y-velocity (km/day)	x-acceleration (km/day^2)	y-acceleration (km/day^2)
Theoretical (initial values):	87.969000	57,910,000.000000	0.000000	0.000000	5,095,600.000000	-468,147.636200	0.000000
dt = 10^{-3} used, one orbit:	87.979000	57,909,999.993600	134.268711	-12.327646	5,095,600.000670	-468,147.627927	50.489716
Absolute error:	0.010000	0.006400	134.268711	12.327646	0.000670	0.008273	50.489716
dt = 10^{-4} used, one orbit:	87.979000	57,909,999.999200	137.723204	-12.652953	5,095,600.000050	-468,147.636161	3.784052
Absolute error:	0.010000	0.000800	137.723204	12.652953	0.000050	0.000039	3.784052
dt = 10^{-5} used, one orbit:	87.978970	57,910,000.000000	-15.110316	1.388230	5,095,600.000000	-468,147.636214	0.672334
Absolute error:	0.009970	0.000000	15.110316	1.388230	0.000000	0.000014	0.672334

Table 1. Values for an orbiting particle (Mercury) around a single mass point (Sun), with basic error analysis of the Euler-Cromer method for one orbital period to test convergence back to initial values used from analytical expressions. In this case, computational times increase by a factor of ten with each decrease in infinitesimal time step, with dt = 10^{-3} requiring ~30 seconds for the code to run. As such, we utilized dt = 10^{-4} to compensate accuracy with computational run time.

The choice to utilize the Euler-Cromer algorithm instead of a more stable, advanced method was multifaceted. First, we didn't study long term stability of our produced orbits and thus didn't encounter significant error accumulation issues. We wanted to properly visualize the changes in an orbit due to changing gravitational conditions, which was evident in just one orbital period. Furthermore, to accomplish these simulations, we needed to construct a functional code which would do the necessary computations of the vector sum of gravitational acceleration at each time step. The well-known 4th-order Runge-Kutta method requires the calculations of multiple coefficients within each time step [4]. Utilizing more advanced differential equation solvers such as this could have slowed code development unnecessarily when considered in addition to producing code which accounted for the summed acceleration from each mass point. Second, given a small enough time step, the Euler-Cromer algorithm produces a sufficiently closed orbit as evident in table 1. It has also been shown to produce similarly accurate results to the Runge-Kutta method with these exceedingly small time steps for short periods of run time [5].

In the literature, others have studied more advanced physical systems of small celestial bodies with nonuniform gravitational fields using spherical harmonics in potential functions, and triaxial ellipsoid or polyhedrons for their geometric setups while presumably using more accurate differential equation solvers [6]. Others have created detailed density and shape models of moons like Phobos which include gravitational information [7]. Our research doesn't delve into these advanced mathematics or topography but does create an elementary way to model mass distribution changes with discretized finite elements while providing a possible teaching mechanism. There are existing interactive physical models for teaching simple Galilean trajectory problems, but not for gravity in such a way where nonuniformities can be easily created and their effects observed and analyzed [8].

3. Results

The creation of our code and simulation system allowed us to model two-dimensional representations of small celestial bodies and orbits around them. We utilized our flood-fill algorithm described previously to populate a grid of some shape, which produced a list of mass point coordinates to be read into our program. We were able to manipulate these lists of values to produce highly irregular shapes and mass-concentrations. As expected, there were clear effects on the orbit as the conditions of the center mass distribution changed. As the complexity of these physical systems increased, so did the sensitivity to the initial conditions and infinitesimal time step for the orbiting particle. We used basic analytical equations for energy conservation, gravitational force, and orbital period to produce guesses for these initial conditions, but had to interpolate on some of these values manually to close the orbit when the mass distributions changed. In particular, the orbiting particle's motion was very sensitive to its initial y-velocity and its x-distance from the surface of the center body. As such, these are the values that showed the clearest changes (along with the orbital period) due to increased mass distribution complexity when other initial conditions were held constant. Figure 1 shows this behavior occurring for a mass distribution only slightly misshapen compared to a large orbital radius and not covering a large spatial extent compared to that of a single mass point. Given the same initial conditions used for one single mass point, we observe that the orbit does not close for the same time interval and would require a slightly higher value for the initial y-velocity to do so.

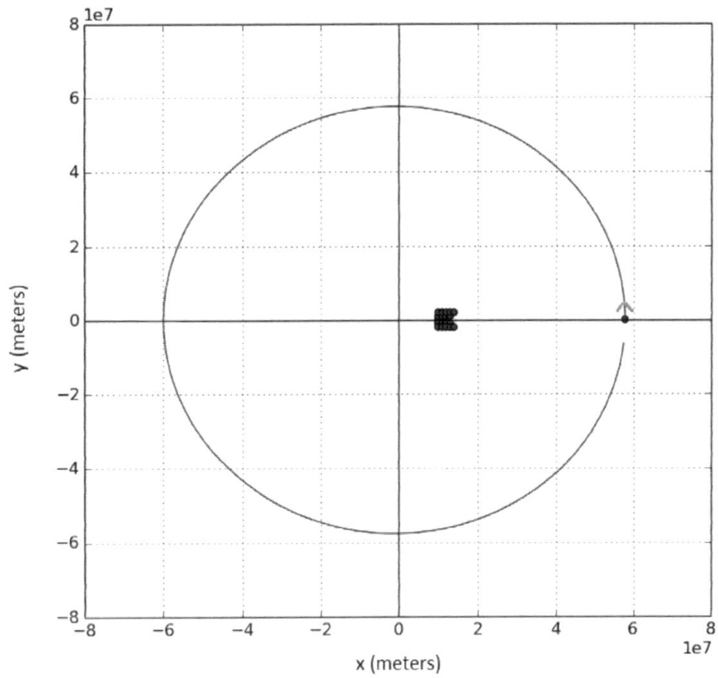

Figure 1. Arbitrary center body shape with 21 center mass points. Using the initial values from table 1 with dt = 10^{-4}, we observe lack of closure of the orbit due to the changing shape of the center body.

In the examples shown in Figure 2, we greatly increase the number of discretized mass points for an ellipse-shaped moon of mass on the order of 10^{22} kg. As we deviate from circular symmetry, the initial values require interpolation as described in the previous paragraph. Moreover, further deviations in mass density distribution also require more extensive interpolation. After discerning the proper initial velocity for the first arbitrary ellipse case on the next page, we manipulated the lists of x and y coordinates of the mass points to include mass-concentrations in asymmetric locations. Without changing the initial velocity before running the simulation, we observe that the orbiting particle crashes into the center body (Figure 2(b)). This is analogous to the issues being studied by NASA's GRAIL missions described in the introduction. Figure 2(c) shows a closure of the orbit around the celestial body with mass-concentrations given the proper initial velocity which was on the order of 10^2 m/s greater than that of figure(a) and figure(b). This illustrates not only the fundamental sensitivity of orbital motion around nonuniform bodies to the initial velocity of the orbiting satellite, but also that these bodies can allow closed orbits around them when the orbiting particle is projected at the correct speed given its distance from the surface.

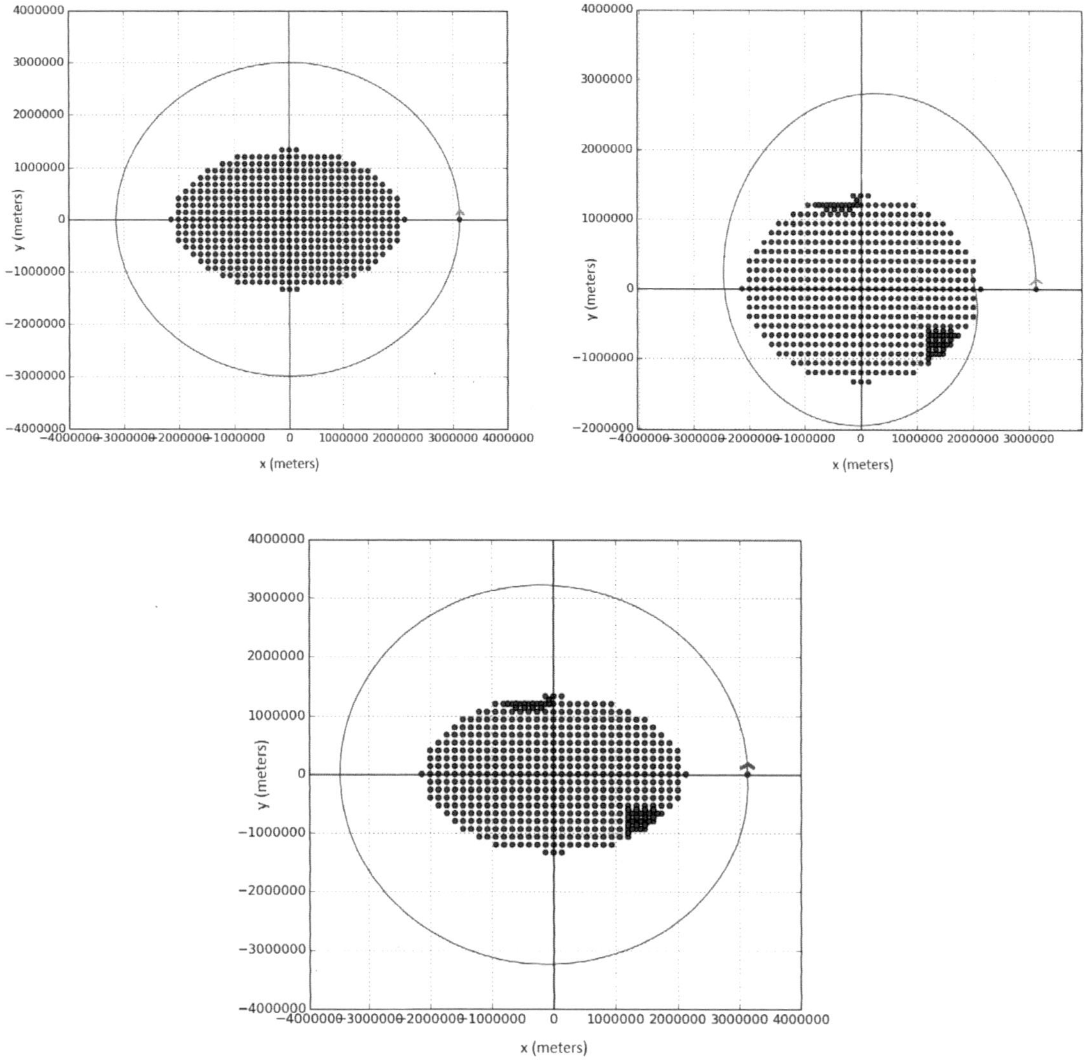

Figure 2. (a) Top left: Closed orbit around uniform moon ellipse shape with 501 center mass points. **(b)** Top right: Attempted orbit around non-uniform body with 629 center mass points, (including "mascons"), with the same initial conditions for the speed and position as in (a), resulting in a crash. **(c)** Bottom: Orbit around same distribution as (b), with adapted initial velocity for a closed orbit.

4. Conclusion

This research succeeded in modeling gravitational anomalies using basic physical properties in a logical but enhanced manner. The technique of summing the gravitational influence of each center mass point in the grid for each time step to represent shape and density changes could be applied in other computational physics settings. This method could also be used as a novel teaching tool to depict gravitational nonuniformities of celestial bodies in an easily understandable manner, using basic Newtonian mechanics and numerical methods.

Furthermore, much can be done to optimize code performance using parallel processing or through employing more accurate differential equation solvers to solve related problems of long term orbital stability and error accumulation reduction. Also, the physical problem could be extended to include other perturbations including orbital inclination in the third dimension, outgassing, or relativity.

Acknowledgments
We would like to thank the Stetson University College of Arts and Sciences Dean's Fund, the Stetson Physics Department, and the Center for Simulational Physics of the University of Georgia for funding.

References

[1] Melosh H, Freed A et al 2013 The origin of lunar mascon basins *Science* **340** 6140 pp 1552-1555
[2] Rossi A, Marzari F and Farinella P 1999 Orbital evolution around irregular bodies *Earth, Planets and Space* **51** 11 pp 1173-1180
[3] Gould H, Tobochnik J and Christian W 1988 *An Introduction to Computer Simulation Methods: Applications to Physical Systems* (Reading: Addison-Wesley)
[4] Newman M 2013 *Computational Physics* (Createspace) chapter 8
[5] Burden R and Faires J 1993 *Numerical Analysis* (Boston: PWS-Kent Publishing)
[6] Cui P and Qiao D 2014 The present status and prospects in the research of orbital dynamics and control near small celestial bodies *Theoretical and Applied Mechanics Letters* **4** 1 013013
[7] Willner K, Shi X and Oberst J. 2014 Phobos' shape and topography models *Planetary and Space Science* **102** pp 51-59
[8] Jimoyiannis A and Komis V 2001 Computer simulations in physics teaching and learning: a case study on students' understanding of trajectory motion *Computers & Education* **36** 2 pp 183-204